Help us make the **future**

Here at Make:, we share stories and projects and create celebrations called Maker Faires because we believe in one simple idea:

We are all makers.

We've brought you Make: magazine since 2005, and more than 1 million people participate in Maker Faires every year across 44 countries.

Together we've built a movement and a global community — but there's so much more to do!

Join us in recognizing the value of the global maker movement and its impact on many, many lives. Your participation will bring the community closer together, help us share more projects and knowledge, and develop new ways to learn.

Become a member today and, together, we can make more makers tomorrow.

Dale Dougherty
Founder of **Make:**

Become a **Make:** member

Learn about the benefits of membership at **make.co**

CONTENTS

Make: **Volume 64** Aug/Sept 2018

CONNECTED EVERYTHING

ON THE COVER:
Jill Ogle with Roxi. Posing assistance from the internet.
Photo: Hep Svadja

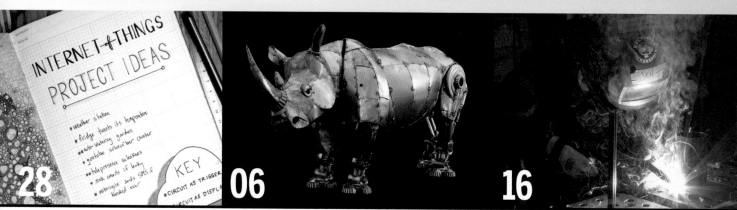

28

06

16

38

42

50

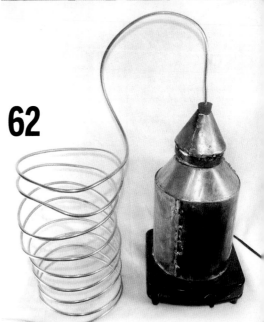

56

62

Make:

> "Sorry, I'm having trouble understanding you right now. Please try a little later."
> —Alexa

EXECUTIVE CHAIRMAN & CEO
Dale Dougherty
dale@makermedia.com

CFO & COO
Todd Sotkiewicz
todd@makermedia.com

EDITORIAL

EDITORIAL DIRECTOR
Roger Stewart
roger@makermedia.com

EXECUTIVE EDITOR
Mike Senese
mike@makermedia.com

SENIOR EDITORS
Keith Hammond
khammond@makermedia.com

Caleb Kraft
caleb@makermedia.com

EDITOR
Laurie Barton

PRODUCTION MANAGER
Craig Couden

BOOKS EDITOR
Patrick Di Justo

CONTRIBUTING EDITORS
William Gurstelle
Charles Platt
Matt Stultz

CONTRIBUTING WRITERS
@blackroomsec, Sam Archer, John Baichtal, Kari Byron, Kat Chapman, Larry Cotton, Tony DiCola, DJ Hard Rich, Tanya Fish, Jen Herchenroeder, Bob Knetzger, Lars Lohn, Brian Lough, Lisa Mecham, Heine Nielsen, Jill Ogle, Jordan Ramée, Joey Ramirez, Michelle Sleeper, Becky Stern, Andrew Stott, Steve Tam, Erik Thorstensson, Greg Treseder, Sarah Vitak

DESIGN, PHOTOGRAPHY & VIDEO

ART DIRECTOR
Juliann Brown

PHOTO EDITOR
Hep Svadja

SENIOR VIDEO PRODUCER
Tyler Winegarner

MAKEZINE.COM

ENGINEERING MANAGER
Jazmine Livingston

WEB/PRODUCT DEVELOPMENT
Rio Roth-Barreiro
Maya Gorton
Pravisti Shrestha
Stephanie Stokes
Alicia Williams

CONTRIBUTING ARTISTS
Bob Knetzger, Jill Ogle, Peter Strain, Mark Taihei, Freepik

ONLINE CONTRIBUTORS
Francisco Sanchez Arroyo, Chuma Asuzu, Kevin Blank, Sarah Boisvert, Gareth Branwyn, Chiara Cechini, Jon Christian, Adam Cohen, David Cole, DC Denison, Shannon Dosemagen, Paloma Fautley, Tanya Fish, Liam Grace-Flood, Grete Kaulinyte, Darrell Maloney, Goli Mohammadi, Violet Su, Adam Woodworth, Kitty Yeung

PARTNERSHIPS & ADVERTISING
makermedia.com/contact-sales or partnerships@makezine.com

DIRECTOR OF PARTNERSHIPS & PROGRAMS
Katie D. Kunde

STRATEGIC PARTNERSHIPS
Cecily Benzon
Brigitte Mullin

DIRECTOR OF MEDIA OPERATIONS
Mara Lincoln

DIGITAL PRODUCT STRATEGY

SENIOR DIRECTOR, CONSUMER EXPERIENCE
Clair Whitmer

MAKER FAIRE

MANAGING DIRECTOR
Sabrina Merlo

MAKER SHARE

DIGITAL COMMUNITY PRODUCT MANAGER
Matthew A. Dalton

COMMERCE

PRODUCT MARKETING MANAGER
Ian Wang

OPERATIONS MANAGER
Rob Bullington

PUBLISHED BY
MAKER MEDIA, INC.
Dale Dougherty

Copyright © 2018 Maker Media, Inc. All rights reserved. Reproduction without permission is prohibited. Printed in the USA by Schumann Printers, Inc.

Comments may be sent to:
editor@makezine.com

Visit us online:
makezine.com

Follow us:
🐦 @make @makerfaire @makershed
google.com/+make
makemagazine
makemagazine
makemagazine
twitch.tv/make
makemagazine

Manage your account online, including change of address: makezine.com/account
866-289-8847 toll-free in U.S. and Canada
818-487-2037,
5 a.m.–5 p.m., PST
cs@readerservices.makezine.com

Building Excitement

BY MIKE SENESE,
executive editor of *Make:* magazine

One of the great things about working at *Make:* is getting to meet creators from all over and see their incredible projects and products. Our latest flagship Maker Faire Bay Area, held this May, was no exception, with an array of cutting-edge technology, stunning art, hands-on exhibits, and so much more. At the event, Arduino unveiled a range of new boards, including the Vidor 4000, their first foray into the FPGA space, promising a drag-and-drop chip-programming interface in the months to come. Oculus invited attendees to try their VR design environment, Medium, and displayed a number of sculptures created virtually and then physically printed out. Solar Rollers demonstrated their sleek, swift educational platform with an engaging racetrack activity. Magnitude.io offered young students the chance to participate in microgravity experiments. And these are just a few of the standouts that wowed the editors — you can see more on page 12 and in our live-updates blog post at makezine.com/2018/05/18/live-updates-maker-faire-bay-area-2018.

We find near-limitless inspiration at this event and the other Maker Faires we get to attend around the world; I strongly recommend checking one — or more — out. When you do, let us know what inspired you the most. With so many amazing things coming from the maker community, we're always happy to hear what our readers get excited about.

Happy making!

CONTRIBUTORS

If it couldn't be hacked, what would you love to connect to the internet and why?

Kari Byron
San Francisco, California
(Bright Ideas)
I struggle to think what isn't connected to the internet already! Maybe a satellite that tracks my daughter in real time. Like a live feed, Big Brother, baby sitter cam. But that is really helicopter mom creepy.

Jill Ogle
San Francisco, California
(Remote Control)
For me it's not about fear of being hacked, it's more about what I have access to hook up. I would love to hook up giant robots, hydraulic presses, or quadcopters if I could. That's what Let's Robot is all about.

Heine Nielsen
Nordjylland, Denmark
(Egg Speakers)
A boat so you always know where it is and you can use the internet if necessary.

Tanya Fish
Sheffield, UK
(Chipped Nails)
I'd like to see the school meals menu connected, so I'd know which days my kid would come home like a ravenous dinosaur and which days I could just give him toast.

Issue No. 64, Aug/Sept 2018. *Make:* (ISSN 1556-2336) is published bimonthly by Maker Media, Inc. in the months of January, March, May, July, September, and November. Maker Media is located at 1700 Montgomery Street, Suite 240, San Francisco, CA 94111. SUBSCRIPTIONS: Send all subscription requests to *Make:*, P.O. Box 17046, North Hollywood, CA 91615-9588 or subscribe online at makezine.com/offer or via phone at (866) 289-8847 (U.S. and Canada); all other countries call (818) 487-2037. Subscriptions are available for $34.99 for 1 year (6 issues) in the United States; in Canada: $39.99 USD; all other countries: $50.09 USD. Periodicals Postage Paid at San Francisco, CA, and at additional mailing offices. POSTMASTER: Send address changes to *Make:*, P.O. Box 17046, North Hollywood, CA 91615-9588. Canada Post Publications Mail Agreement Number 41129568. CANADA POSTMASTER: Send address changes to: Maker Media, PO Box 456, Niagara Falls, ON L2E 6V2

PRINTED WITH
SOY INK

Robots and Representation

Hep Svadja

Judy Gilmore
@Judy_Gilmore_   Follow

Daughter reading @make magazine this Saturday morning. Glad to see females well represented in the magazine! She's inspired!

10:31 AM - 14 Apr 2018

3 Retweets 20 Likes

♡ 1   ⟲ 3   ♡ 20

Pravin Vaz
@pravin_vaz   Follow

Electronics outreach using @make easy #electronics by Charles Platt at @chanelcollege went really well. Students want another round next term making some advanced projects.

3:12 PM - 8 May 2018 from Masterton, Masterton District

6 Retweets 13 Likes

♡ 1   ⟲ 6   ♡ 13

MOVED TO MAKE INMOOV

When I was 10 years old, I went to my first Maker Faire and it was very inspiring! Afterward, I started working with an Arduino kit, making my own circuits, and trying to modify the code with my mom. It was a summer later when I was truly inspired, I read a *Make:* magazine with an article about the world's first open source 3D printed humanoid robot called an InMoov. The best part was there were ordinary people all over the world making their own InMoov. I knew that if they could make an adult-sized humanoid, then I could make the robot just as well.

In the beginning of the summer when I was 12, my mother bought a LutzBot Mini 3D printer. I found a file of the InMoov head on Thingiverse and it took around a month to print and build. Making the torso took the rest of the summer, where I had plenty of time to do whatever I wanted in those long summer days. Making the robot's torso quickly became a habit, but the best part was when mom came home and saw what I had accomplished throughout the day.

We are looking forward to another summer of making robot upgrades!

–*Lurene Davis, age 14, via email*

SPIES LIKE US

I have to tell you that (Issue 62) came at the perfect time. My 6-year-old daughter really wants to be a secret spy! I actually ended up pulling out a few old solder learning kits I had in the toolbox (FM radio, FM bug transmitter, and super ears) that we just finished putting together. Next is the coffee cup camera featured in the mag and a modified version of the laser trip wire! Thanks for the focus on women in DIY and tech, I love showing my daughter that there are role models out there who didn't get where they are because of what they looked like. –*Matt Leshko, via email*

CYBERPUNK COUNTRY

I look at *Make:* magazine for inspiration and information. I liked that the cyberpunk issue featured projects I can afford to be inspired by — I don't need to be in a higher tax bracket to afford the tools to make the stuff.

I also liked the breakaway to an actual subculture. I don't know any level 80 night elves, but I do know a whole bunch of cyberpunks, unlicensed radio enthusiasts, and backwoods rural Maine hackers. It was nice to see a little of that in *Make:*.

–*Eric Lovejoy, via email*

HAVE SOMETHING TO SAY?
We want to hear!
Send us your stories, photos, gripes, and successes to
editor@makezine.com

MADE
ON EARTH

Backyard builds from around the globe

Know a project that would be perfect for Made on Earth?
Let us know: *makezine.com/contribute*

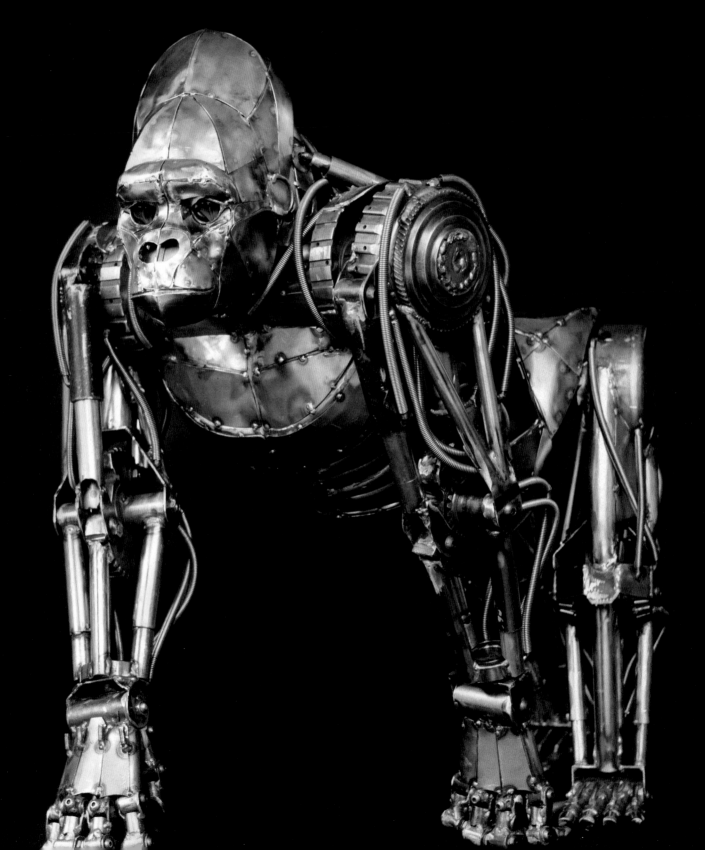

METAL MENAGERIE

ANDREWCHASE.COM

Sometimes you make a seemingly mundane decision that alters the course of your life forever. For **Andrew Chase** that decision was purchasing a small welder to create metal security grates for the windows of his photography studio. But Chase caught the welding bug. He started welding all sorts of furniture. Everything from beds and tables to couches and chairs. Despite being borderline unsittable, the furniture took off and a number of people commissioned pieces from him.

Chase made his first animal, a giraffe, to be a character in a picture book. Because the character would need to be in different positions for each part of the story he opted to add posable joints to its body.

Chase says he enjoys "making the sculptures articulated ... partly because doing so gives the pieces a sense of realism and honesty that they otherwise might not have." But getting the pieces just right is quite a process. Chase starts each by overlaying an image of a skeleton on a side-view image of the animal. This helps him to position the joints and get the proportions correct for the first step, which is a crude skeleton made only of tubing, joints, and ¼-inch rod. Finally he fleshes out the whole piece with 20-gauge sheet metal.

One of the hardest parts is "figuring out what to leave in and what to leave out," according to Chase. "Generally speaking, the more I fill in the less movement I get so I have to leave gaps and voids so the joints have freedom to move. More detail equals less articulation," he says.

Almost all of the metal Chase uses in his pieces is recycled, with much of it coming free from auto shops. Ultimately, his works end up weighing between 55 pounds (cheetah) to 150 pounds (rhino). Chase usually keeps the original and uses it as a template to make more animals.

When asked what's next he says, "I'm thinking about doing a Chinese dragon. I've avoided mythical creatures so far but it feels like it's time for a change." —*Sarah Vitak*

Andrew Chase

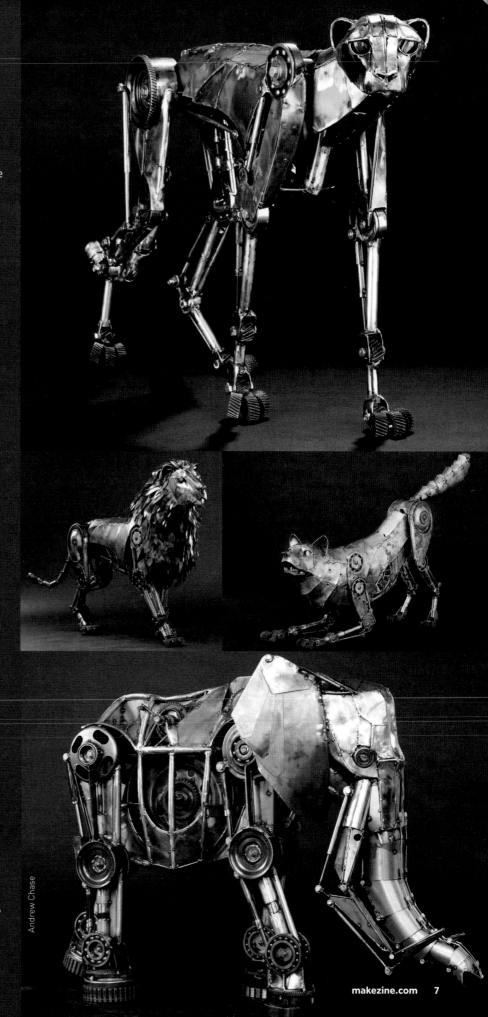

VIRTUAL VISAGE

MATTHEWMOHR.COM

If you're planning a trip to the Greater Columbus Convention Center, you'll want to check out the 14-foot human head in the North Atrium. *As We Are* is a massive construction of LED screens that display a person's head at roughly 17 times its normal size. Visitors who want to participate in the exhibit step inside a photo booth that's hidden in the center of the screens. The booth takes a 3D photo of the participant's head and then *As We Are* transforms into a larger-than-life depiction. The giant portraits encourage onlookers to draw their own conclusions as to each person's identity and history.

Matthew Mohr, a professor at the Columbus College of Art and Design who's made a career out of utilizing technology in service of art, is behind the concept. "I want people coming to Columbus, Ohio to feel what I feel about the city," he says. "[Columbus] is a forward-thinking, increasingly diverse community where everyone is welcome to engage and contribute. *As We Are* is that first point of engagement."

Although an extensive system of engineering work and digital fabrication is responsible for the exhibit's fantastical transformation from one face into another, Mohr is happy to see most people react to the display as art and not as technology. "I've seen many participants have a mini-existential crisis," Mohr says. "In an age of near constant visual exposure to manufactured images, experiential art has a true opportunity to connect and communicate."

And Mohr isn't done yet. "[*As We Are*] is the first of its kind in many ways and, if I might add, this is only the beginning," Mohr says. "There is added functionality that will come to light in the next couple of years."
—Jordan Ramée

Matthew Mohr

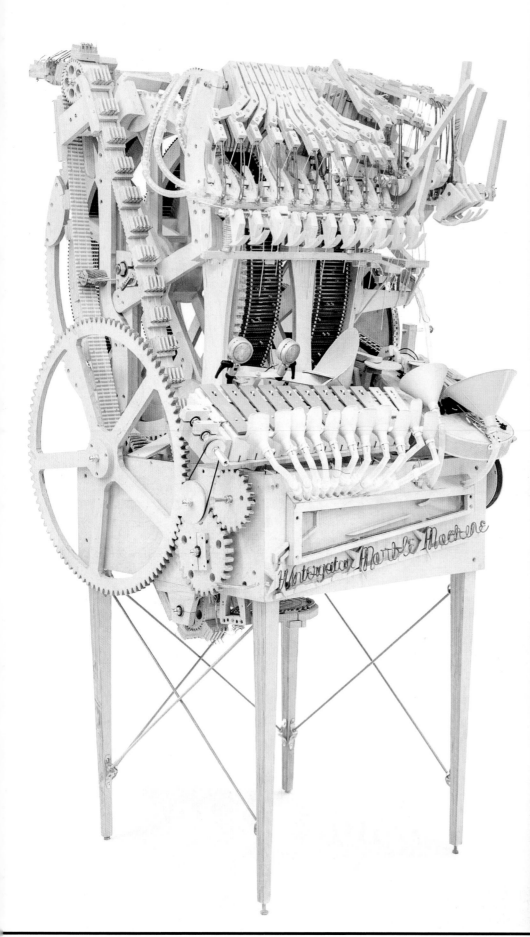

MARBLE MADNESS

WINTERGATAN.NET

You've no doubt seen the video of Swedish music group Wintergatan's Marble Machine: a crank-driven plywood apparatus, standing taller than most people, that creates music by dropping more than 2,000 metal marbles onto various musical surfaces like the bars of a xylophone, strings of a bass guitar, or drum pads.

The Marble Machine was created primarily by band member **Martin Molin**. He was inspired to build it after visiting the Museum Speelklok — a gallery for self-playing musical machines, in Utrecht, the Netherlands.

Whether it's the programmable percussion wheel, the divider that separates the marbles into each individual music track, or designing the housing for the contact mics so the snare drum sounds like a snare drum and not a bass kick, every single mechanism, every note, every instrument on the Marble Machine was another opportunity to solve a problem. Now Molin is tackling an even bigger problem: Building a sturdier, more reliable version of the Marble Machine to take on tour — the Marble Machine X.

Fortunately, Martin isn't alone this time. He's put together a development team to help him build Marble Machine X, and he's got the whole internet on his side too — online fans of the project are constantly helping out by suggesting new mechanisms or methods of engineering to bring the Marble Machine X to life.

It'd be easy to write it off as just a crazy machine that makes beautiful music, but it's so much more than that. The Marble Machine is an engine for music that is fuelled by ingenuity, creativity, community, engineering, tenacity, and heart.
—*Tyler Winegarner*

Wintergatan

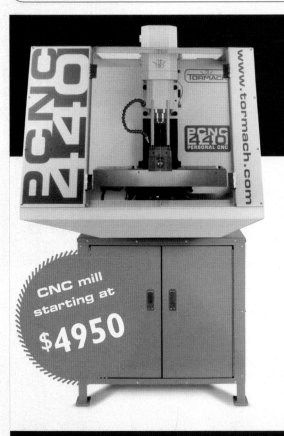

MAKER FAIRE MAGIC

Becca Henry, Jun Shéna

1 BioLumin0ids
This beautiful, ethereal glass sculpture by Erick Dunn is light- and sound-animated, as well as interactive: The computer code can be altered to change the animations in real time.

2 Sharky-Go-Round
Blacksmith and shark enthusiast Kirk McNeill created this alluring cooperative, interactive spiral of 10 hammerhead sharks to bring awareness to stopping the practice of shark finning. Children (or adults!) working together can spin the three-ton base.

3 Cosmic Space Worm
Take in the scene with your friends on this segmented multi-rider bicycle with over 500 color-changing LEDs, created by Tyler FuQua.

4 Introspection
This piece by artist James Peterson provides a peaceful, light-filled place to take a rest amid the hustle and bustle of Maker Faire.

5 Real Life Spaceteam
The team at Particle built this physical version of the popular cooperative mobile game *Spaceteam* using seven Particle Photons, a lot of FeatherWings, and custom adaptors. Seven players attempt to repair their failing spaceship with seven briefcases containing real hardware views of the ship systems.

6 Dream Pipes
Clody Cates' massive interactive PVC water sculpture includes smoke effects, calming water sounds, and iridescent rainbow animations to create a soothing and inspiring atmosphere.

7 Clock Ship Tere
Andy Tibbetts fabricated this 33-foot-tall engineering masterpiece, which includes a hubless front wheel and sails made of fire.

8 Rabid Transit
Constructed from recycled scrap metal by Duane Flatmo and Jerry Kunkel, the result is a fantastical robotic pyrotechnic extravaganza.

9 Prosthesis
Mixing form and function, this real life mechanical exoskeleton is the first human-controlled racing mech.

Bright Ideas

MythBuster Kari Byron and "geek mom" Debra Ansell chat about technology, art, and the magic of being self-taught

Written by Kari Byron

In her new book *Crash Test Girl: An Unlikely Experiment in Using the Scientific Method to Answer Life's Toughest Questions*, *MythBuster* Kari Byron presents all facets of her background as science experiments, stepping bravely through her education, relationships, career, and more, analyzing the lessons learned in each part and giving smart advice to readers based on the results. It's surprisingly honest, highly useful, and totally hilarious all at once.
—*Mike Senese, executive editor,* Make:

I FEEL LIKE I WAS BORN A "MAKER." As an artsy kid, I was always busy constructing something. I even lived in a giant cardboard box rocket ship in my living room for as long as my parents could tolerate it. If you know my origin story (I love that comic book expression) and how I became part of *MythBusters*, you know I wanted to be a model maker and get into special effects. I sought out an internship at Jamie Hyneman's M5 Industries so I could continue my love of making. I never realized there were so many out there just like me until *MythBusters* really caught its stride. That was right around the time I started to hear about "makers" and *Make:* magazine.

Every now and then I meet a maker that really impresses me, usually with a skill set opposite of mine. Technophile mom Debra Ansell of GeekMomProjects.com is just that. At a past Maker Faire I coveted her Twitter-enabled LED handbag. Nothing I love more than someone who is brilliant *and* creative! I started to internet-stalk her so I could ask about *her* origin story and find out what else she has in her glowing bag of tricks.

Kari: Were you like me, a maker even as a kid?
Debra: As a child I had a lot of project ideas, but was always very frustrated by the difference in the way I would imagine them versus the way they would actually turn out. I was (still am) clumsy and not very artistic. I think that much of my interest in making things now has been driven by computers and cheap microcontrollers which allow me to execute instructions with precision, as well as the availability of CAD and tools like 3D printers and laser cutters which make it so easy to transform an idea into a concrete object. These days, if a project doesn't turn out as I expected, I have the ability to keep tweaking the specifications until it does. It's such a satisfying process.

I hope to copy your headband project and impress my daughter. As a cyber native she will be a natural at a programmable DIY craft like that. How did it come about?
I was looking for a programmable wearable project that could be completed in a few hours by people without any special technical skills (sewing or soldering). Knowing the possibilities of the newest tiny microcontrollers, I kept mulling over ideas. My first idea was a handbag. The headband idea just evolved from that. I was forced to do a very small amount of soldering for the project, and create a super-simple PCB to connect the LEDs to the microcontroller, but other than that it is easily assembled with off-the-shelf parts.

How complicated can the light sequence get?
Because the CircuitPython LED code generator is drag and drop, there is a limit on the complexity of the patterns generated, though I've tried to make individual code blocks that represent relatively complex functions, e.g. "twinkling" the lights randomly or scrolling a phrase across the headband in Morse code. From what I've read the size of the Circuit Python code file is limited to 30–40KB, or about 250 lines of code. I've run into that limitation a couple of times and tried to compensate by reusing as much code as possible. There are also limitations on the amount of RAM available, so my code generator "Brightly" isn't really good for programming long strings of lights (the headband has 14 LEDs, which is pretty close to the number you can have and still use my code generator to specify interesting patterns). The next generation of chips that will run Circuit Python are supposed to solve this problem with more memory and RAM, so I'm not going to try too hard to work around it at this point — I'll just wait and let hardware solve the problem. You can still generate code for many different interesting patterns within the current limitations.

Your adaptation sounds like fun to play with.
I'm planning to make "Brightly" generally available — you can currently see it in action here: brightwearables.com/brightly/index.html. Drag and drop the code blocks from the menus on the left side of the webpage into the workspace to make the program, and then click the "Download Code" icon to download a CircuitPython file called *main.py*. I just want to tweak it a bit more, and hopefully add a bit of explanation before officially releasing it.

How'd you learn to program Python?
I took a year of programming in college (LISP and C), but I'm mostly self-taught, and as a result, I have a lot of bad programming habits. I first coded regularly when I was in physics graduate school at Cornell and used FORTRAN to evaluate my data files. That really dates me! I actually was employed as a software engineer for a while after graduate school, but I had no real official training, and cringe internally when I look back at my code from that time period. I tend to learn code on an as-needed basis for my various projects. I taught myself Python when I built a V-plotter because I wanted to create a cross-platform GUI interface, which Python does well. I also chose Python because it has a ton of built-in modules that handle complex image manipulation and large data arrays, which I needed for the plotter. It's a really interesting language, and ridiculously powerful. I still don't feel like I know it well, though.

It is hard to find fun tech projects that appeal to my daughter. I hear that sentiment all the time from my other mom friends.
I'm hopeful that the headband project is fun and accessible enough that it will appeal to girls who aren't as excited by robotics or other tech projects. Programming your own clothing and accessories is a fun and unusual way to learn to code, and I hope it engages a wide audience of curious makers who might not have been motivated by other kinds of project. ◯

Debra Ansell rocking her LED headband

Headband guts: Gemma M0 and an LED strip

Debra's gorgeous light-up handbag

LED accents work nicely with denim

See more from Debra at
GeekMomProjects.com

Written by Jen Herchenroeder

Fast Food

BattleBots team Poor Life Choices cooked up their
Battle Royale with Cheese in just five weeks

HAVE YOU EVER HIT ON AN IDEA SO GOOD THE WHOLE ROOM BURSTS INTO LAUGHTER? Enter the cheeseburger bot …

Our *BattleBots* group Poor Life Choices was born of a desire to build something the world had never seen. Half the team was harboring childhood dreams of building for *BattleBots*, the other half was just crazy enough to try to build a combat cheeseburger for the fun of it.

Once we settled on the basic burger design we had five weeks to move from first draft to finished robot. As a first-year team thrown into a field of seasoned, multi-decade veterans, we knew we had tremendous challenges to overcome. Mistakes were made

and friendships were tested. A lot of late night coffees and burgers were consumed. Many favors were asked.

A FAB TEAM
From the start Miles Pekala worked furiously to CAD up the burger components into plans that could be shared with our small groups — all Power Racing Series alums — in three cities: Oakland, Baltimore, and Chicago. In Oakland, I led the workshop at NIMBY Space where Jordan Bunker and Lindsay Oliver joined in fabrication and logistics. This group worked as the core of creation and assembly, with frequent field trips to local metal suppliers and hackerspaces.

Our teammates Charles Wittington, Brice Farrell, Jeremy Ashinghurst, and Angela Rothbaum were back east and connected by Baltimore Hackerspace. This group consulted on and machined the parts for the weapon system and kept an eye on the overall budget, asking favors at The Foundery for machining aluminum components and seeking sources of funding at every level. Of all the roles you can have on a build, team accountant is often the unsung hero.

CHALLENGES
One of the major fabrication issues was creating and aligning the pieces for the interior tube frame. Twenty-foot sections

Team Poor Life Choices

(From left)
Charles Wittington
Jordan Bunker
Lindsay Oliver
Jen Herchenroeder
Miles Pekala
Jim Burke

(*Not pictured*)
Brice Farrell
Jeremy Ashinghurst
Angela Rothbaum)

JEN HERCHENROEDER is an applied technology specialist and visual artist based in Oakland, California.

---

of .75" OD (outer diameter) A513 tube steel were acquired, cut, and carried over to industrial art space M0xy and its Department of Spontaneous Combustion, where a very old and suspiciously safety-less tube roller lives in the back of the metal shop. With no emergency stop or user protection at all, we named this contraption "Bendytube Crushyourthumbs."

This tool didn't have the appropriate dies for the patty, which was created from 2" OD structural tube steel. For that we had to ask for a favor from Zach Wetzel at his workshop in Richmond, California.

Back at NIMBY we needed welding fixtures for nearly every joint type. Precision was crucial for notching and aligning a few hundred feet of carefully bent tube. Jordan and I worked at this for many sleepless nights, salvaging wood and MDF and creating CAD and CAM files to cut the

fixtures on a little Shapeoko CNC.

The toughest obstacle on the entire robot came from needing to run the two systems — drive and weapon — on one battery. The drive is running two brushed DC motors, MY1020 style, the kind you'd buy for electric scooters. The weapon is run on a brushless RC airplane motor, the Toro Beast. We couldn't have integrated these to run on the same 10s LiPo battery without the help of our friends.

BRINGING HOME THE BACON

We needed a bacon blade, and Advanced Metalcraft in Chicago has a laser cutter powerful enough for hardened steel. Luckily, our honorary teammate, Jakub Klimuszko, a member of Chicago's Pumping Station: One makerspace (which was founded by teammates Jordan Bunker and Jim Burke),

is an accomplished welder and fabricator working for Advanced Metalcraft. He lobbied his boss Peter Anwar and the company agreed to sponsor us with a very generous donation of materials. Jakub then donated over 100 hours of his own time to brake-bend and weld-up the buns of steel!

Despite the time crunch, a great amount of research went into bearings, shafts, and couplers to get the bacon swinging. Due to the shape of the armor, getting the weapon and drive to fit together and work together without interference was a real challenge. The clearance between the three motors was so tight that we had to shave down the bolts on the drive motors to get the weapon in. We were definitely living dangerously as we approached the deadline!

COMPLICATIONS

Of course, there were unexpected hiccups. Fitting the gorgeous armor to the tube frame we realized that the frame was curved on a radius that matched the armor version one, but what we'd sent out in CAD and received back in steel was armor version two. Whoops. Hacking happened.

We were also missing some key parts when we hit the road: The vendor we purchased the 10s LiPo batteries from delayed shipping so long that they didn't show up in Oakland until after we had left for Long Beach to film. We had to find emergency LiPos that would fit in the space allotted under the motors, and we had to make do with 8s LiPos. The day we arrived back home the correct batteries were waiting by the door, taunting us.

HOORAY FOR HELP

When we began the project we had no sponsorship at all. By a few strokes of excellent fortune, friends at DigiKey and Imgur received word of our crazy little burger robot and contributed crucial funds to the build. A local burger joint in San Francisco, WesBurger, also joined in. We really couldn't have gone the distance without this help!

Despite the challenges, the time, and the complications, we got that burger to the arena! 10/10 would build again. As I write this new episodes are airing weekly on Discovery Channel and, having seen all those robots up close, I have to say you're in for a great season. ⊘

Written by Jordan Ramée

Flying Toward the Future

The Daedalus suit soars closer to realizing our jetpack dreams

THOUGH PLENTY OF AIRCRAFT HAVE ALLOWED US TO SOAR AMONG THE CLOUDS, we're still determined to build a future where a trip through the sky is as simple as stepping into a rocket-powered suit. Now, thanks to Richard Browning's Daedalus, this future might not be all that far off. Much like the ancient Greek craftsman it's named after, this propulsion suit bestows the power of flight.

Browning started work on Daedalus back in 2016, and showcased the suit's first published test flight live at a TED conference one year later. "[The suit] came from a deep respect for the capability of the human mind and body, and was inspired by my time with the Royal Marines Reserve, as well as taking part in triathlons, ultra-running, and calisthenics," Browning says.

JORDAN RAMÉE spends most of his time writing about geek culture. Although he's particularly passionate about game design and Japanese art, he loves traveling around the world to meet creators from all walks of life.

A REAL-LIFE TONY STARK

Browning's startup, **Gravity** (gravity.co), is responsible for continuing work on the suit. The company is still tinkering with the Daedalus Mark 3, but Browning has high hopes for the suit's immediate future. Filmmakers, who want to step away from CGI and rely on practical effects, have already approached him about buying the suit.

When movies come up, Browning's Daedalus is almost always immediately compared to Tony Stark's Iron Man suit. "Interestingly enough," Browning says, "the first *Iron Man* film, where Tony Stark is shown learning to fly, demonstrates just how good the CGI folks were in thinking about and understanding the physics. They portrayed something very close to what we've learned in practice."

UNDER THE ARMOR

The current suit relies on six jet engines (each of which can provide up to 22kg of thrust) to fly. The engines on the suit's arms control direction and speed, while a display inside the helmet lists updates on fuel consumption and the pilot's current altitude. Browning is already expanding on the suit's technology and working to include a more expansive HUD system, a wireless data link to a ground station, and an airbag.

Browning admits the original suit "required a lot of core and arm strength," but goes on to say he and his team "have engineered [the new suit] to rely less and less on strength, so that any moderately fit individual can fly." Despite those reassurances, the suit still looks incredibly difficult to use. In videos of Browning flying the suit, you can see his arms shaking in response to the constant pressure of maintaining his balance.

In one such video, the Daedalus nabbed the record for fastest speed in a body-controlled jet engine power suit (32 miles per hour). Browning assures us the suit can go a whole lot faster, and higher too. He says he and his crew are just being careful and purposively holding back — though the Daedalus itself can currently take quite a hit, the human body inside the suit wouldn't fare as well during a crash. "It's less about armor and more about impact absorption," Browning said, "and ahead of that, it's about rapid and safe transition to parachute height, which we are close to achieving." ◗

More Advances in Personal Human Flight

HOVERBIKE
hover-bike.com
Malloy Aeronautics' most recent project, the Hoverbike, might be its coolest drone yet. Designed to fly at the same speed and altitude as a small helicopter, the Hoverbike is also capable of operating near the ground around people. The controls are intuitive, so pilots won't have to undergo rigorous training to fly one. Capable of lifting 286 pounds, it can seat a pilot or be flown remotely.

EHANG 184 AUTONOMOUS ARIEL VEHICLE
ehang.com/ehang184
Currently only coming to Dubai, the EHang 184 is able to carry a single passenger weighing up to 250 pounds for 30 minutes. It's pretty simple to operate too: Just climb into the cockpit, input a destination on the control panel, and then relax as the 184 flies to your chosen coordinates. The drone flies at 62mph, and is monitored remotely from a control room. Safety measures ensure that it won't take off in poor weather and will land immediately, and safely, if problems occur.

FLYBOARD AIR — ZAPATA RACING
zapata.com/air-products/flyboardair
The Flyboard Air stays afloat through the power of four 250 horsepower engines. A logic system inside the hoverboard, similar to the one found in drones, helps stabilize the machine — however, the pilot still needs to make constant adjustments while in the air. Franky Zapata, a professional French jet ski pilot and founder of Zapata Racing, says flying the Flyboard Air will probably take 50-100 hours of practice.

SCORPION — HOVERSURF
hoversurf.com
Russian-American aero design company Hoversurf has successfully tested a flying four-rotor manned drone, called the Scorpion, which can be piloted like a motorcycle. The company is giving the design to law enforcement in Dubai. Scorpion pilots, adorned with body armor and flying at frighteningly fast speeds, look a lot like *Star Wars* scout troopers.

CONNECTED EVERYTHING

Our smart devices and projects will make our lives easier —
as long as we are as smart as they are **Written by Mike Senese**

Security once simply meant locking your deadbolt and keeping your porch lights on. Our tech-infused lives, however, now bring us a number of considerations that would have been impossible to imagine even a couple decades ago.

The hard part about digital security and privacy is that it's so broad. There's hardly a part of our lives that doesn't have a connected aspect to it now, and that shows no sign of slowing down. Cameras, microphones, sensors, and sensitive data are all intertwined with the great promise of convenience and efficiency — and in more ways than we probably realize, it's working. When Waze routes you around a bad patch of traffic, we celebrate gaining a few extra minutes. Telling Siri or Alexa to

set a reminder for you can keep you from missing an important call or from getting a pricey parking ticket. And using an IP camera to peek at your sleeping baby can bring a needed sense of home when on a work trip far away.

But these tools have created opportunities for digital infractions that we are still only starting to understand. Some of them are user generated (my toddler yells "yes!" to anything Alexa asks, including confirmation that I want to order something); some are flaws of the system (such as Alexa recently sending a couple's recorded conversation to a random contact); and many others are as malicious as they are clever (one casino had its high-roller database stolen by hackers who broke into

the system through its aquarium's internet-connected thermometer last year).

The good news is there are likely things you can do to upgrade the security of your connected devices and DIY IoT builds. In the following pages we'll look at a connected-device search engine that can find unpatched connected gadgets (reminder: do frequent updates!), review best practices for your home IoT builds, and discuss some of the initiatives that various companies are pursuing to enable increased privacy and security in our digital lives. We'll also guide you through using a fun, fast telepresence robot system and leveraging IFTTT to get started on smart, wireless projects. Go ahead and get connected — just be smart about it.

JILL OGLE makes robots controlled by the internet. She's been obsessed with this idea since leaving her cushy game designer job at Disney Interactive, and set out to create a platform that everyone could use — letsrobot.tv.

REMOTE CONTROL
Let's Robot offers an open telepresence platform for everyone   Written by Jill Ogle

Let's Robot is a site where anyone from anywhere can take control of real life robots via the internet. The robots stream audio and video in real time with little to no latency. There is a chat room that you can use to talk to the other people sharing the experience, or you can use it to make the robots talk to the humans on the other side.

Anyone can actually add their own robot. Since we opened the site last year, more than 700 different homebrew robots have gone live on the site from over 200 different creators around the world.

The interface can be fully customized for each robot, allowing you to add simple wheeled robots, or complicated ones with multiple degrees of freedom. When things get busy, there is a live dynamic voting system that allows the crowd to decide what collective action to take next in real time.

Each Robocaster (this is what we call our broadcasters) has a page they can customize, tell people about their robot builds, and special rules they want their users to follow. We even sell a couple of kits in our store if you want to stream, but don't know anything about robots. Our website is still an early prototype, so there are still a lot of kinks to work out.

Television Tested

One of our more popular robocasters goes by the name Roboempress. Her creations include Kame, a modified turtlebot with a French accent, Killer Krawler, an awesome robotic ATV designed for outdoors, and PP Cannon, an air-powered high velocity ping-pong cannon. Roboempress allows users to shoot her with ping-pong balls while she works.

Last year, *The Chris Gethard Show*, which airs live on Tru TV, contacted us about having live-controlled robots on their show for an episode. We commissioned Roboempress to make another ping-pong cannon for them, and made a couple of smaller robots to run around their office. We shipped everything to New York just

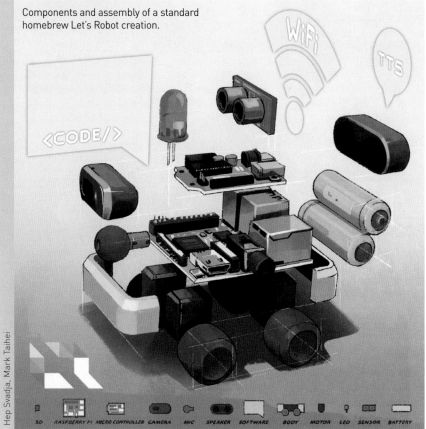

Components and assembly of a standard homebrew Let's Robot creation.

Hep Svadja, Mark Taihei

in time for the episode "Technology will destroy us all." We were surprised to find out their guest for that evening was none other than John Oliver. Our site had trouble handling the rush of traffic, but thankfully we were able to pelt Mr. Oliver in the face a few times during the episode live on air. They also had a gong that shut off all the show's lights if hit. Even with many users sharing and voting on controls, everyone was able to organize beyond expectations and become very disruptive for their episode.

Connecting Your Own Robot

If you're building your own, start here: letsrobot.readme.io/docs/building-your-robot. The robot side software is open source, and runs on most Linux-based computers. We even have an API that allows you to fully customize the experience: letsrobot.readme.io/v1.0/reference.

Most of the homebrew robots on Let's Robot use the following components:

- Raspberry Pi or other single-board computer. The newest Raspberry Pi has onboard Wi-Fi, you just need to point it at your access point.
- SD card with Raspbian or NOOBS installed. You can follow our guide to get our software to run on your robot, and pair it with the site: letsrobot.tv/setup.
- Microcontroller, such as Arduino. The Adafruit motor hat is also popular.
- Camera to see
- Microphone to hear
- Speaker to let the robot talk
- Body to hold all the parts
- Motors and servos to move and drive around
- LEDs and sensors to make things interesting
- And a battery to power it all

A lot of devices and robots are already supported by our software, including the GoPiGo Robot, and Anki Cozmo. If you have an awesome robot just sitting on the shelf collecting some dust, this could be a great way to share it with everyone!

If all of this sounds like too much, we also sell a development kit called "Telly Bot," which works out of the box with the letsrobot.tv site. We'll see you online. ◉

Awesome user-created robots

1. **Trc202** created a lawn mower robot, and another that plays PlayStation games. You can also sometimes play with his cat.

2. **Datadrian** converted an old power wheels vehicle to hang out with him at the beach.

3. Grabby never sleeps, and neither does its tireless creator, **Mikey**.

4. **Banzai Baby** is a partnered Twitch Streamer who also lets her viewers control the robot in her mad scientist workshop.

5. **Dope250** has a small fleet, Madriva, Stanley Bot (pictured), and Stacey Bot.

6. **Opkillie** made this robot out of freaking wood!

7. Keep the party spinning with **Nocturnal**'s fidget spinner bot.

8. Look out for **mbrumlow**'s spider bot, it is on the prowl.

9. **Chad** often takes Bottington out on adventures to the park!

10. **Jose** likes to chill with his big bots, like Fantabulosa.

11. Everyone agrees, **Jill**'s Roxi is the cutest!

12. Flem is a happy bot! Created by **Clem**

13. How about a dose of Furby bot by **Fatmeatball**? We are still waiting for a Furby organ!

14. **Spoon**'s claw bot has all the claw action you are looking for!

SEEK AND DEPLOY

How hackers use the **Shodan search engine** to discover and take advantage of your connected devices

Written by @blackroomsec

The growth of the connected device world is matched by stories of these tools and gadgets being exploited by hackers and bots. Rarely, however, do the stories describe the processes used to find and take advantage of our networked appliances. One of the more common and powerful tools used for this is called Shodan.

Shodan is the go-to search engine we hackers typically use when hunting for devices that are internet-facing and which have services running on open ports that may be exploitable, depending on a variety of factors. The device does not always have to be an IoT device like an IP camera but can be a full-fledged server, computer, or even a virtual computer. For the purposes of this very simple demonstration, I am going to illustrate two searches we can use on Shodan to look for devices with special conditions applied and explain the hacker way of thinking behind it all.

Example 1: SLMail

The first step is to find devices that have vulnerable services. I did a search on Shodan for devices that are running SLMail, version 5.5.0.4433 which is vulnerable to a Remote Buffer Overflow exploitable condition as described in cvedetails.com/cve/CVE-2003-0264. A CVE (Common Vulnerabilities & Exposures) is a technical document explaining the vulnerability in more detail, the version(s) of the software it affects, and in some cases, whether any efforts have been made to fix the issue with a new version or some other method of mitigation and if exploits are available for it publicly or in Metasploit modules.

> Metasploit is a hacking framework tool that assists in the creation of exploits for software vulnerabilities. It also has scanners in it that allow the user to search devices to see if they are vulnerable to specific conditions and vulnerabilities themselves.

This particular SLMail vulnerability allows anyone to remotely exploit the server and take control over it as if they were the admin or root user provided they have version 5.5.0.4433 installed. They do not need to

be authenticated with a password nor do they need to be on the same network as the vulnerable computer. The search term I used was simply "slmail" (Figure A).

I have highlighted the vulnerable server in red. As this vulnerability in this software is well known among hackers and there are many public exploits available, someone with very little technical or hacking knowledge could exploit this and gain control over the system. I have redacted some of the info like the IP address shown on the screen. Armed with the IP address of a vulnerable server, the hacker may now take any of the public exploits available for this, edit it to fit the conditions of the computer they are going to be using to attack, and launch the attack against the remote server.

Example 2: Hikvision Camera

Hikvision is a manufacturer of many popular types of IP cameras. In September of 2017 there was a public disclosure of what is known as a Privilege-Escalation vulnerability in many Hikvision cameras, seen at seclists.org/fulldisclosure/2017/Sep/23.

Privilege Escalation means to escalate or elevate one's current access to higher access with the goal being the admin or root user. Because the article doesn't list the specific models which contain this backdoor, the hacker will need to get creative in trying to find public cameras which may have it installed by using logic.

The author mentions that starting in January of 2017, Hikvision began to mitigate this issue and the backdoors were removed with updated software patches to these devices. So let's search for "hikvision 2015" since the backdoors were known to have existed in that year (Figure B).

In order to see if these particular cameras in the search results are vulnerable we will need to visit the IP address of the camera (replacing IPADDRESS with the actual IP) with the following URL as mentioned in the article:
IPADDRESS/Security/
users?auth=YWRtaW46MTEK

If it's vulnerable, it will show us a list of users on the device. This is not the only function of the backdoor and the article should be read fully to understand what else is possible.

Patch Your Software

Unpatched software in a device that is publicly accessible is dangerous because it tends to give the impression to hackers that they should stick around and see what else they can discover and try to hack. It indicates that the operators of these devices are not actively updating their equipment and are quite possibly not aware that they should. We can then take this inference and search their network by using the term net: (e.g. net:"IPADDRESS.0/24") with the appropriate CIDR (Classless Inter-Domain Routing) notation and see if there are any other devices listed in the database that may also be vulnerable to other kinds of attack vectors.

Many IoT devices, along with public-facing servers, contain vulnerabilities that are not patched to prevent unauthorized access to them despite heavy media focus on this topic.

Specific Searches

Shodan is an extremely robust tool and many other search terms can be used to find results more in line with what you are searching for, if one should be interested in taking this further. ⊘

@BLACKROOMSEC
BlackRoomSec is a white hat hacker and professional pentester from N.Y. who runs a blog aimed at teaching new learners and encouraging them that "hacking is not a hobby but a way of life."

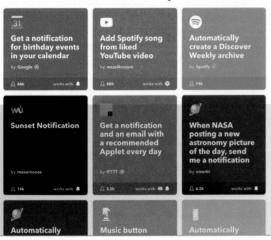

IF THIS THEN THAT

Get started with the popular IFTTT cloud service for gluing together web services and hardware   Written by Becky Stern

*I*f it's going to rain tomorrow, *then* light up my umbrella stand. *If* my Fitbit recognizes that I'm awake, *then* start brewing coffee. *If* my cat uses his door, *then* log the date and time in a Google spreadsheet.

If This Then That, also called IFTTT, is a site that aggregates and provides an interface for a wide variety of web services and smart devices, which all have Application Program Interfaces, or APIs — published parameters for how software programs should interact with one another. Advanced software developers may be able to whip up a web interface for their DIY electronics project in a weekend, but for the rest of us, IFTTT is an API gateway that handles much of the heavy lifting, code-wise, to easily create internet-connected hardware projects with complex interactivity.

What's IFTTT Good For?

The primary focus of the site is service-to-service, like updating your phone's volume settings when you leave and return home, or automating your Instagram photos to post natively to Twitter, or syncing your Spotify and SoundCloud. Sometimes these smart triggers work so seamlessly that you forget they're even there.

Beyond automating your own device settings and web services, and logging your own data, commercial IoT device representation on IFTTT has grown quickly in recent years. So in addition to your Philips Hue lights and WeMo devices, you can now also connect your Google or Amazon virtual assistant, your GE, LG, Whirlpool, or Samsung washer, dryer, dishwasher, and fridge, your smart window blinds, your thermostat, your car — at the time of this writing, IFTTT has eight services just for gardening.

For DIYers, IFTTT makes it incredibly easy to whip up electronics projects that ping you via email or SMS text when, say, the water heater is leaking. It's similarly straightforward to design your own projects that receive data from the web, like a weather station.

In my design grad class, students have 15 weeks to create functional prototypes of their product designs using Arduino and IFTTT. Most have never touched electronics or coding before, and by semester's end they never fail to impress me with how sophisticated their prototypes' interactions are.

How to Use IFTTT

Head to IFTTT.com and click the "Sign up" button to create an account. While you're at it, install the IFTTT app if you've got an iOS or Android device (Figure A). Two-factor authentication is highly recommended, as you will be linking up your personal accounts on services like Twitter, Amazon Alexa, Nest, your email, etc. Keep your account safe from hackers and spambots by using a unique password and enabling two-factor authentication to confirm your account by SMS.

Now, here's how to use it:

1. Link up your accounts on any services you already actively use. This gives IFTTT permission to access your accounts through the APIs.

2. Experiment with popular recipes for the devices and services you've connected. Familiarizing yourself with the applet-creating process now will make it easier to integrate your hardware projects later.

3. Brainstorm a list of project ideas (Figure B), then categorize them. Is the circuit triggering an internet action, acting as a display for web info, or both? It's a bit simpler to start with a project that detects a physical input and sends a trigger to the web, then work your way up to more features.

WiFi Weather Display by Becky Stern,
instructables.com/id/WiFi-Weather-
Display-With-ESP8266

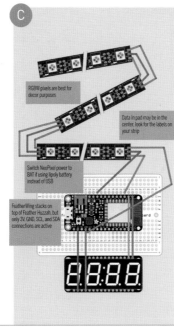

RGBW pixels are best for decor purposes

Data in pad may be in the center, look for the labels on your strip

Switch NeoPixel power to BAT if using lilpoly battery instead of USB

FeatherWing stacks on top of Feather Huzzah, but only 3V, GND, SCL, and SDA connections are active

Monitor a feed on Adafruit IO

This Trigger fires anytime it validates the data that you send to your feed. Example: If Feed Temperature > 80, fire Trigger.

Feed (required)

command

The name of the feed to check.

Relationship (required)

equal to

Relationship between two values.

Value (required)

1

The value to compare against.

Create trigger

TIP: Don't make the common mistake of biting off more than you can chew at one time. Always develop hardware projects by testing out individual features/components before combining them.

4. Pick some internet-connected hardware to get started on your first project. If you already know Arduino, you might consider an ESP8266 or ESP32 board such as NodeMCU or Adafruit Huzzah. Or maybe you'll opt to start with a Raspberry Pi. IFTTT also specifically supports hardware by Particle, littleBits, Seeed, and a handful of "single button" devices you can find if you scroll through the expanded list of services on IFTTT's site.

5. Work up a circuit diagram showing connections between your main controller and any outputs (LEDs, seven-segment displays, etc.) and inputs (sensors) (Figure C).

6. Get on a Wi-Fi network with clean access to the outside web (no captive portal used by school/business networks, for example).

7. Set up your hardware. Depending on what hardware you're using, you may also need to install additional software and/or libraries and sign up for additional cloud data services sites like adafruit.io, Particle Device Cloud, and Amazon Web Services (AWS). These sites work in combination with your hardware's project code to read and write values to your own custom feeds. Then the cloud service's IFTTT app can access those custom feeds to integrate with the rest of the site (Figure D).

Now you're ready to write your own IFTTT recipes to make your IoT project do your bidding. ⊘

[+] Learn more about using IFTTT in your DIY electronics projects by completing my free Instructables Internet of Things class, which uses Arduino, ESP8266, adafruit.io, and IFTTT together. instructables.com/class/Internet-of-Things-Class

Primo Prototypes Pronto

Here are a few examples of projects made by my students to help spark your creativity and demonstrate the power to build prototypes fast with IFTTT. They're all published on Instructables, too: instructables.com/id/Making-Studio-at-SVA-PoD

① Carly Simmons' **Social Circle Relationship Manager** tracks how often you text with cherished friends and family, and lights up LEDs to remind you to keep up with those little interactions that add up to a lot. instructables.com/id/Social-Circle-Relationship-Manager

② Will Crum wanted litter-box harmony in his home, so he built a **Smart Litter Box Sensor** to detect ammonia and alert him and his girlfriend that it's time to scoop, rather than relying on the partner with the more sensitive nose. instructables.com/id/DIY-Smart-Litter-Box-Sensor

③ Rhea Bhandari's **Angry Aunty** stops procrastination by detecting when you enter your home and promptly calling you on the phone to remind you to get straight to your homework before you have a chance to crash out on the couch, either with an automated message or by hitting up your actual angry aunty with a message to call you. instructables.com/id/Angry-Aunty-Comes-to-the-Rescue

④ Smruti Adya's mom is always checking in on her from halfway around the world: "Where are you? And have you eaten?" Wanting to go about her busy NYC grad student life, Smruti built her mom a **Location and Food Clock** that tracks her phone and displays whether she's at school, at home, or "other," and displays the status of her meals with LEDs. Now mom can rest easy and Smruti's got some well-deserved space. instructables.com/id/Location-and-Food-Clock

⑤ Josh Corn created **Blüp, the Bubble Notifier**. It's a sleek glass cylinder wired up with an air pump to release a bubble that slowly rises through the liquid (hand soap) to subtly gain your attention — "to not only notify me of an event of my choosing, but to also use the time it takes for the bubble to rise to instill a sense of urgency." instructables.com/id/Blüp-the-Bubble-Notifier

Becky Stern, Carly Simmons, Will Crum, Rhea Bhandari, Smruti Adya, Josh Corn

LOCK IT DOWN

Top tips to **tighten security** on your homebrew IoT projects

Written by Tony DiCola and Brian Lough

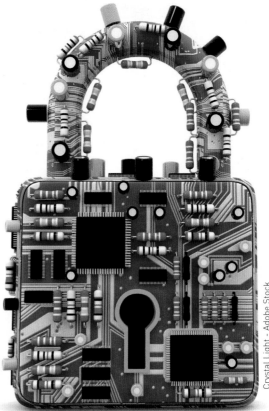

Crystal Light - Adobe Stock

For Raspberry Pi / Linux Boards

1. Change the default user password.

All Raspbian OS installs use the same password for the default **pi** user. Attackers will try common passwords first. Stop them by using the **passwd** command to change yours to a new, unique, and strong value.

2. Disable password login with SSH.

Set up *security keys* to login to your board instead. These keys grant access only from other allowed computers, using sophisticated cryptography that's extremely difficult to guess or break. And you won't have to remember passwords anymore!

3. Keep the OS up to date.

Inevitably, vulnerabilities and bugs will be found. Diligently use the **apt-get upgrade** command to update your board's operating system with the latest security fixes.

4. Set up a Pi firewall.

Your IoT gadget doesn't use all the services the Pi operating system provides, such as web server, email server, and more. Use the **ufw** tool to enable a firewall on your board. Turn off all services by default, then only turn on those you're using.

5. Review Linux security best practices.

DigitalOcean's guides "Introduction to Securing Your Linux VPS" and "7 Security Measures to Protect Your Servers" explain tools like **fail2ban** and **tripwire** that help detect intrusions and deter attackers.

For All IoT Devices

1. Change all default passwords.

It's the number one vulnerability for internet-connected devices. Your router, IP cameras, network printers — if you can look up the password online, so can everyone else!

2. Keep firmware/software up to date.

Stay protected from known vulnerabilities.

3. Disable services and protocols you don't need.

If your device isn't using SSH or RDP or FTP, disable them. Every way of connecting to a device is a vulnerability.

4. Only expose to the internet what you need to expose.

Your router's firewall should block access to devices on your internal network — unless you configured it otherwise. To give yourself access to devices on your network, you may have configured *port forwarding* on your router. It's important to enable this *only for secure devices*.

5. Use a VPN.

A *virtual private network* is like creating a secure tunnel into your home network that's not exposed to the internet. You can use your laptop or phone to connect securely to your device from anywhere.

6. Obscurity is not security!

Haven't shared a link to your device? So what. There are bots that literally scan the internet looking for devices to exploit. Anything that's exposed to the internet should have authentication on it. Enable passwords for web interfaces if they're available.

7. Use a guest network.

Set up a guest network on your router for your IoT devices. Even if one of your devices is breached, your main network should remain secure.

8. Try a third-party message broker

such as Adafruit.io or even Telegram messenger. These can be a more secure way of communicating with your device. ●

BRIAN LOUGH is a software developer who got into Arduino development after discovering the ESP8266 chip.

TONY DiCOLA is a software engineer experienced in cloud services and embedded systems such as Arduino, RaspPi, and MicroPython.

[+] Read and share the longer version of this article, with more tips and links to great security resources, at makezine.com/go/iot-sec-tips.

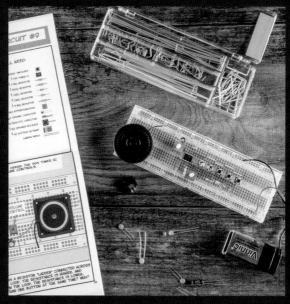

Kits by
Make:

**get hands-on and
build your DIY skills**

makershed.com

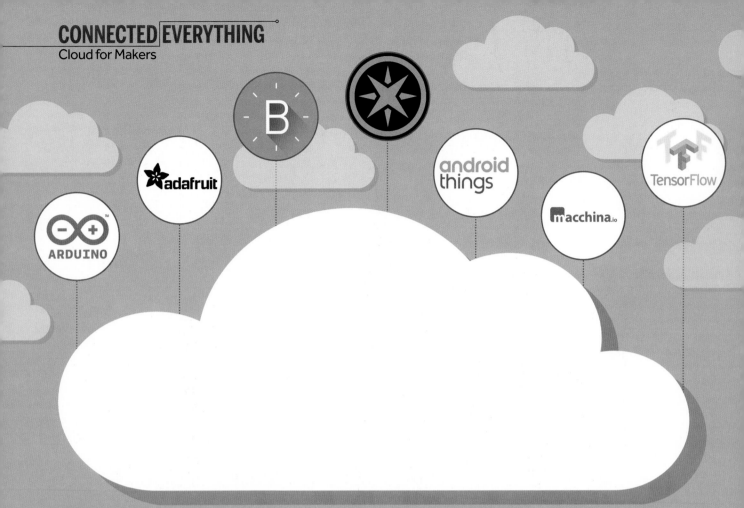

AT YOUR SERVICE

Take your IoT project online with these maker-friendly options **Written by Hep Svadja**

With our connected world rapidly expanding, several cloud services have sprung up to cater to the IoT maker. Depending on your preferred development environment or language, they offer a choice of services to suit every IoT need. Most include a REST API which allows easy access to control or communicate with hardware, and several have access to SDKs for building your own apps to interact with, display, and interpret your data.

Arduino Cloud
cloud.arduino.cc

Many makers don't realize that Arduino's online development environment also includes access to cloud services. This allows you to develop your code and set your cloud interactions, as well as find cloud-specific libraries. Arduino Cloud

was engineered to work seamlessly with Arduino's line of IoT-friendly boards, such as the MKR1000, and includes a pretty intuitive setup to link your connected devices to the cloud libraries and services you want to access. Bonus: It also lets you program Raspberry Pi and BeagleBone.

Adafruit Cloud
io.adafruit.com

Adafruit's cloud service provides a user-friendly UI experience along with their traditional easy-to-follow learning guides and a variety of interesting examples.

HEP SVADJA is *Make:*'s photographer and photo editor. In her spare time she is a space enthusiast, metal fabricator, and *Godzilla* fangirl.

Adafruit.io is great for novice IoT makers, but also strong enough to keep up with those who are more experienced. With access via REST API in any language that speaks HTTP, this means you can program in the language that feels comfortable to you.

Blynk
blynk.cc

Blynk is an IoT service that lets you interact with devices from your phone without having to code your own native app. It is hardware agnostic with tons of supported components, and if your particular board isn't included there is a wide array of add-on shields, hats, and capes that can provide access. The Blynk Code Builder allows you to assemble your sketches easily, or you can use the Arduino IDE by installing the Blynk libraries.

Particle Device Cloud

particle.io/products/software/device-cloud

The data-gateway Device Cloud is fantastic for those who have invested in the Particle platform. Particle's new Mesh technology makes this an attractive choice for those who need to deploy devices in areas with limited Wi-Fi or cellular coverage. Device Cloud has both desktop and web IDE access, along with SDKs for both web and mobile app development, and a REST API which makes it easy to interact with the boards.

Android Things

developers.google.com/iot

Android Things allows you to leverage Google tools such as machine learning libraries and Google Assistant into your IoT devices. It supports single-board computers such as Raspberry Pi and System on Modules such as the Qualcomm SDA624. Android Things is scale-ready as well, if you plan on supporting hundreds of devices.

Macchina

macchina.io

Another high-level option for embedded Linux systems geared towards maker pros, Macchina is a powerful development platform. It works with both new and legacy code bases, allowing you to leverage the power of C++ from JavaScript, without having to write connecting code. Applications can be containerized, making it easy to deploy and manage third-party applications in a secure framework. Developers have the choice of free open source or commercially licensed accounts. The latter can be pricey, but if you're looking to deploy on a production level, it's a decent option that is robust enough to handle what you throw at it.

TensorFlow Lite

tensorflow.org

TensorFlow Lite is designed for deep neural network models, but in a smaller binary size than original TensorFlow, with fewer dependencies, decreasing the memory load on small chips. TensorFlow Lite also lets you hit the hardware accelerator through the Android Neural Networks API, which is designed for running computationally intensive machine learning operations on mobile devices. While developed for the advanced user, TensorFlow also offers a lot of user education materials on getting started with machine learning basics, as well as detailed tutorials for deepening your AI skills. ○

Get Off My Cloud
Keeping your private data closer to home

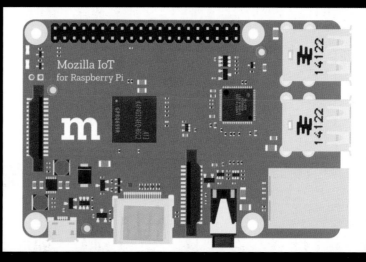

Mozilla Project Things

iot.mozilla.org

The Things Gateway is a smart home hub you can build yourself with a Raspberry Pi, which allows you to directly monitor and control your home over the web without a middleman. It works with smart home devices from a range of different brands, using various protocols, to talk to each other by using the web as a common layer. The gateway hosts a web interface to control all your devices, lay them out on a floor plan of your home, create "if this then that" style rules to automate them, and even control them with your voice. This all happens on your local network so your private data stays inside your home by default, and it can continue working locally even without an internet connection. —*Lars Lohn*

Snips

makers.snips.ai

Snips is an AI voice company that has put a big focus on solving the problem of companies "listening in" on your home. The company has just launched a nifty Maker Kit, which runs on a Raspberry Pi — completely on-device and private by design. This means you can control devices like your Sonos speakers, Hue lights, and more, with no internet connection and your data never leaving your home. —*Steve Tam*

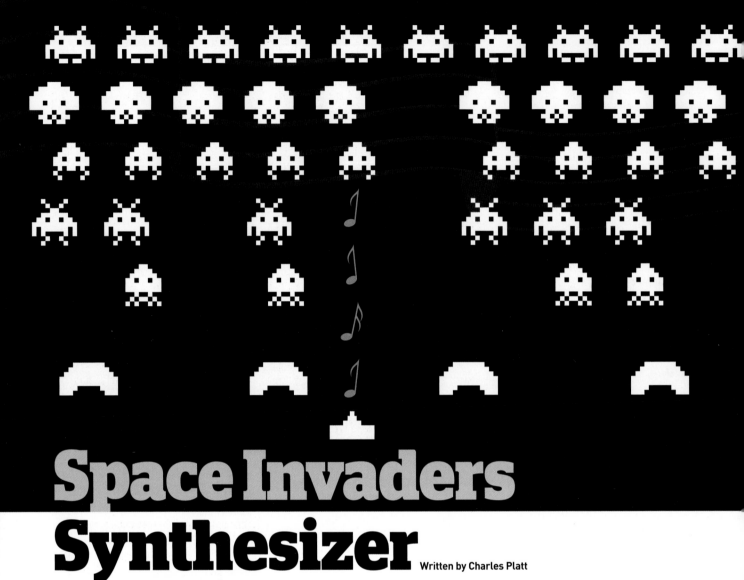

Space Invaders
Synthesizer

Written by Charles Platt

Hack the original 1978 arcade audio chip for classic and novel sound FX

CHARLES PLATT is the author of *Make: Electronics*, an introductory guide for all ages, its sequel *Make: More Electronics*, and the 3-volume *Encyclopedia of Electronic Components*. His new book, *Make: Tools*, is available now. makershed.com/platt

Forty years ago, the SN76477 was born: a legendary chip designed by Texas Instruments to create sound effects in arcade games. If you've ever played *Space Invaders*, you've listened to the SN76477. A sample is shown in Figure Ⓐ.

In those primitive times, controlling this chip was a challenge for hobbyists. You could use rotary switches and pushbuttons (as I'll suggest below), but if you got ambitious and hoped, maybe, you could run it from the cassette port on an Apple II — well, maybe not.

Everything is now much easier, because many functions of the SN76477 are set via logic inputs. Just send high or low signals from an Arduino or any other 5V

microcontroller, and you can sequence a cacophony of rifle shots, sirens, the puffing of a steam locomotive, and a sentimental lullaby. Better still, all of the sounds will share that retro arcade ambience.

Despite the age and obsolescence of the SN76477, you can still buy samples from multiple eBay suppliers for around $15. Apply a 9V battery, and the chip contains its own voltage regulator which converts the power to 5V internally and also makes it available (up to 10mA) from a 5V output. This you can use to power the logic inputs.

Sound from the chip has to be amplified, and the manufacturer recommends using a couple of transistors. I found that a single 2N3904 would work.

TIME REQUIRED:
4–5 Hours

DIFFICULTY:
Intermediate

COST:
$20–$40

MATERIALS

» **SN76477 sound generator IC chip** about $15 on eBay
» **Solderless breadboards (3)**
» **Capacitors:** 100pF (1), 150pF (1), 300pF (1), 500pF (2), 1nF (1), 10nF (1), 22nF (1), 50nF (2), 68nF (1), 100nF (2), 220nF (1), 470nF (3), 1µF (3), 10µF (4), and 50µF (1)
» **Resistors:** 100Ω (1), 7.5kΩ (6), 22kΩ (1), 47kΩ (3), 50kΩ (2), 100kΩ (2), 220kΩ (1), 330kΩ (1), 1MΩ (1), and 10MΩ (1)
» **Trimmer potentiometers:** 50kΩ (2) and 1MΩ (6)
» **Switches:** SPST slide (9), SPST momentary pushbutton (1), SPDT (7), 5-position rotary (7)
» **Transistor, NPN, 2N3904**
» **Loudspeaker, 8Ω**
» **Jumper wires, assorted**
» **Battery, 9V**
» **Snap connector for 9V battery** with leads

TOOLS
» **Wire strippers / cutters** if you're making your own jumper wires

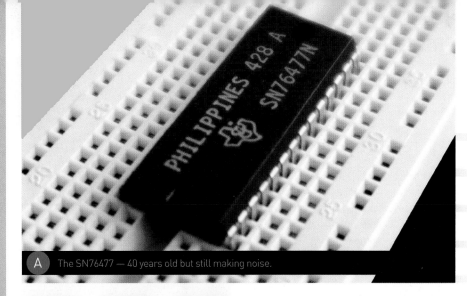

A The SN76477 — 40 years old but still making noise.

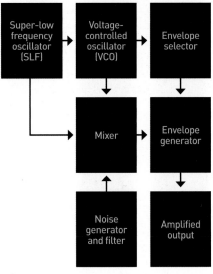

B Simplified diagram of the sound chip's internal features.

A scan of the original datasheet is kindly maintained by Experimentalists Anonymous at experimentalistsanonymous. com/diy/Datasheets/SN76477.pdf. Before playing with the SN76477, you should download a copy. Careful study of this seminal document showed me that the chip synthesizes sound from three sources inside itself, as shown in Figure **B**. A voltage-controlled oscillator (VCO) creates tones that vary in pitch with voltage, and a super-low frequency oscillator (SLF) can control the VCO, creating the irritating *whoop-whoop* or *weoo-weoo-weoo* sounds that will induce paroxysms of nostalgia in arcade dwellers of the 1980s. An additional noise generator creates white noise, which is useful to simulate explosions.

A mixer can blend any two or all three of the sources, and an envelope generator can modify the attack and decay of one-shot sounds such as a bird chirping.

The mixer merges sounds by ANDing them. If you want to hear sounds that can be distinguished from each other even though they occur simultaneously (for instance, bird song during a nuclear explosion) you have to alternate the mixer inputs at a rate of around 50kHz. This can be done with a 555 timer and a multiplexer.

All the sound attributes are adjustable with external resistors and capacitors. If you prefer to select capacitors using a microcontroller rather than rotary switches, you'll need more multiplexers. I don't have space to get into that topic here, but if you have my book *Make: More Electronics*, it describes multiplexers in some detail. You may also consider using digital potentiometers anywhere you see potentiometers in the schematic, because digital pots are specifically designed to be microcontroller-friendly.

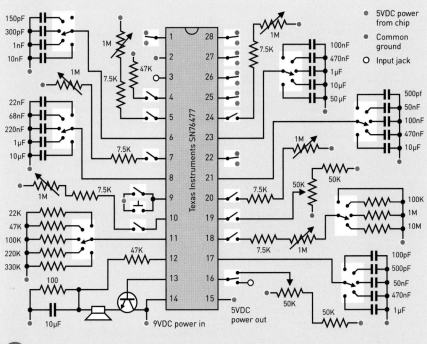

C Test circuit derived from a schematic by Texas Instruments.

Figure ⓒ on the previous page shows a test circuit that I derived from a version helpfully supplied by Texas Instruments when the chip was first introduced. This is not as complicated as it may seem. To breadboard it, I used three single-bus boards side by side, as shown in Figure ⓓ. Instead of rotary switches, I used movable jumper wires. Figure ⓔ shows a label that you can cut out and stick on the chip to keep track of the pin numbering.

So now if you've been putting off the chore of synthesizing the sound of a crashing car while an electronic organ plays "Twinkle, Twinkle, Little Star," you have no further excuses. The SN76477 can do it all. ⊘

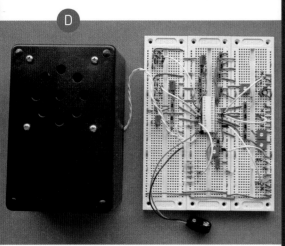

The large number of selectable capacitors and resistors will be most easily managed using three breadboards.

ⓔ

1	28
2	27
3	26
4	25
5	24
6	23
7	22
8	21
9	20
10	19
11	18
12	17
13	16
14	15

You can copy, print, and paste this label on your chip to keep track of the pin numbers.

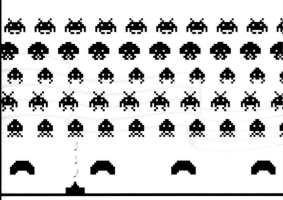

SN76477 PIN FUNCTIONS AND COMPONENT VALUES

In Figure ⓒ, I have retained most of the original component values suggested by Texas Instruments, but you can substitute standard values that are similar. Switched capacitors are used to establish a broad range for each key feature, while potentiometers make adjustments within each range.

1. Logic input. Applies a sound envelope in conjunction with pin 28.

2. Negative ground.

3. External clock input for noise generator (optional). Maximum 10V.

4. Closing the switch enables the internal noise clock. The 47K resistor may be increased up to 100K to get a lower-frequency noise.

5. Low-pass noise filter adjustment.

6. Low-pass noise filter range.

7. Decay activation switch and adjustment. Determines the discharge time of the attack/decay range capacitor applied to pin 8. Open switch causes instant-off (no decay time).

8. Attack/decay range.

9. Logic input. Logic high inhibits sound output, logic low (or open switch) enables sound, and high-low transition with the pushbutton triggers the one-shot function. If the input goes high during a sound, the sound will be interrupted. Envelope selection pin 1 must be high and pin 28 must be low for the one-shot to work.

10. Attack activation switch and adjustment. Determines the charge time of the attack/decay range capacitor applied to pin 8. Open switch causes instant-on (no attack time).

11. Audio output level.

12. Feedback from amplifier output.

13. Amplifier output to base of transistor.

14. 9VDC power input. Also supplies collector of transistor.

15. 5VDC power input if pin 14 is unconnected, or 5VDC power output if 9VDC is applied to pin 14.

16. External VCO input, or internal VCO adjustment. Maximum range of external source: 0V to 2.35V. Exceeding this range saturates the audio output and causes distortion.

17. VCO range.

18. VCO activation switch and adjustment.

19. Pitch adjustment via pulse-width modulation of the VCO output. Leaving the switch open gives a 50% duty cycle of the VCO.

20. Super-low-frequency oscillator activation switch and adjustment.

21. Super-low-frequency oscillator range.

22. Logic input. Logic high controls VCO with internal capacitors, logic low selects external control with pin 16.

23. One-shot duration range.

24. One-shot duration adjustment.

25. Logic input. Logic high selects Mixer B.

26. Logic input. Logic high selects Mixer A.

27. Logic input. Logic high selects Mixer C.

28. Logic input. Applies a sound envelope in conjunction with pin 1.

The result of combining the logic states of pins 25, 26, and 27 to select mixer inputs is shown in Figure ⓕ.

The result of combining the logic states of pins 1 and 28 to select sound envelopes is shown in Figure ⓖ. For example, if pin 1 is low and pin 28 is high, the mixer output is fed continuously to the amplifier without any envelope being applied.

ⓕ

Pin states			Mixer output
Pin 25	Pin 26	Pin 27	
○	○	○	V
○	○	●	S
○	●	○	N
○	●	●	V+N
●	○	○	S+N
●	○	●	V+S+N
●	●	○	V+S
●	●	●	Inhibit

○ Logic low   ● Logic high

V = Voltage-controlled oscillator
S = Super-low-frequency oscillator
N = Noise generator

The mixing of sound sources is controlled by the logic states of pins 25, 26, and 27.

ⓖ

Pin states		Envelope applied to
Pin 1	Pin 28	
○	○	V
○	●	Continuous
●	○	One-shot
●	●	Alternating

○ Logic low   ● Logic high

V = Voltage-controlled oscillator (VCO)
Continuous = No sound envelope
One-shot = Triggered by pin 9
Alternating = Alternating sound cycles applied by VCO

After setting an envelope for your sound (i.e. attack, sustain, and decay) you decide how to apply it by setting logic states of pins 1 and 28.

Charles Platt

Powerslide

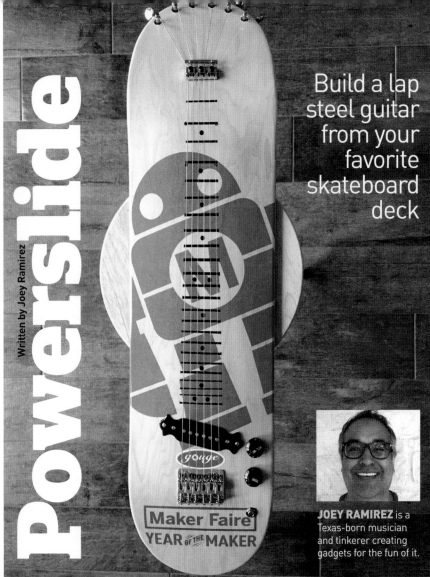

Written by Joey Ramirez

Build a lap steel guitar from your favorite skateboard deck

TIME REQUIRED:
2–4 Hours

DIFFICULTY:
Intermediate

COST:
$50–$100

MATERIALS

- » **Skateboard** Doesn't have to be new but if it is used, make sure it's solid.
- » **Guitar bridge**
- » **Guitar nut**
- » **Tuners (6)**
- » **Pickup with pots and input jack**
- » **Fret measurement printout**
- » **Wood screws (5)** for bridge and nut

Below will vary with above parts:
- » **Stainless steel washers, 1/8"×1" (2)**
- » **Stainless steel machine screws, #6-32×1" (2)** cut to length

TOOLS

- » **Drill and/or drill press**
- » **Drill bits**
- » **Spade bit, Forstner bit, or hole saw**
- » **Chisel**
- » **Hammer**
- » **Screwdriver**
- » **Punch or nail**
- » **Ruler or tape measure**
- » **Painter's tape**
- » **Dry erase marker or wax pencil** so you can remove your marks

JOEY RAMIREZ is a Texas-born musician and tinkerer creating gadgets for the fun of it.

Joey Ramirez

 A
 B
 C
 D
 E
 F

To start off, I can't take full credit — I saw similar skateboard guitars online but never found a step by step. I am not a guitar builder so your suggestions are welcome. You can find more details at instructables.com/id/Lap-Slide-Guitar-Skate-Board.

1. INSTALL TUNERS

Select one end of the deck for your tuners. Measure and mark the placement of the holes, and then drill them to size. I was too lazy to get my drill press out, so my holes chipped a bit. Install the tuners (Figure **A**).

2. INSTALL BRIDGE

Mark out where you want the bridge (I goofed on my first attempt — the sticker covers those holes). I used two of the holes for the trucks (the rear ones), and added washers to make it more stable (Figure **B**). Cut the bolts to length so they don't stick out the back and poke your legs.

3. INSTALL NUT

Measure 20½" from the end of the bridge and mark where to drill. Drill the holes and screw the nut down (Figure **C**).

4. CUT PICKUP HOLE

Mark where you want the pickup, then surround it with painter's tape to prevent scratching the board when cutting. I used a jigsaw, then a "scroll" blade for the tight turns (Figure **D**). Do not mount the pickup yet.

5. PAINT FRETBOARD

Using the PDF found on Instructables, apply painter's tape to make the measured "fret" marks on your fretless guitar. Take your time, and use a check-off list to mark which frets you've completed. Once taped, cover the rest of the board so you don't get off-spray on it (Figure **E**) and apply spray paint — I did one coat. Wait 30 minutes and carefully peel off the painter's tape.

6. INSTALL PICKUP

Mark where you want your pots and drill out some of the wood with a Forstner or spade bit. I went about halfway through the board to get the nut side of the pot to have some thread to tighten the bolt. I used a chisel to take out some wood for the lugs of the pots, but you could always make the holes wider instead. Screw down the pots and pickup and you're good to go (Figure **F**). Your guitar is now ready to be stringed! ⊘

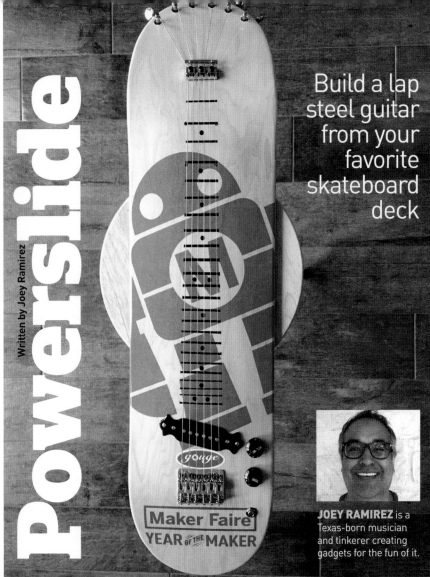

Synthetic Sounds

Written by
DJ Hard Rich

Make a minimal sliding synth and drum machine for just a few bucks each

Here are two portable yet capable electronic instruments I've designed that can be built for $15 to $20. The Syntheslider (Figure Ⓐ) is a monophonic digital synth that you manipulate with your thumbs; the Drum Machine (Figure Ⓑ) is a four-channel pushbutton beat maker. They're a ton of fun.

I use STM32 "Blue Pill" boards for each, which can be had for under $3 and are clocked at a blistering 72MHz with 64kB (or 128kB in many cases) of flash memory — great for storing samples and offering signal processing options.

Both projects also use the Mozzi sound synthesis library by Tim Barrass. It's extremely fast and flexible, can output via digital pin or I2S DAC, and is compatible with all AVR Arduinos. Mozzi includes functions for digital synthesis, sample playback/manipulation, as well as DSP effects such as reverb or delay. Tim's done a great job creating a slew of other examples and documentation; check out the Mozzi homepage at sensorium.github. io/Mozzi and see what features you can add to these projects.

BUILD YOUR SYNTHESLIDER AND DRUM MACHINE

For each build, start by preparing the board and libraries — most of these steps are the same for each build, with the input type (sliders or buttons) and Mozzi library settings being the main differences.

First, wire the breadboard following the diagram for the Syntheslider (Figure Ⓒ) or the Drum Machine (Figure Ⓓ). Solder the header pins to the STM32 if the board does not already have headers.

Next, connect the ST-LINK programmer to the 4 SPI pins opposite the micro-USB connector on the STM32. The markings are under the pins — match GND, 3.3V, SWDIO, and SWCLK to their respective pins on the ST-LINK.

Then install STM32 support in the Arduino IDE by following the instructions at github.

DJ HARD RICH is an audio hacker from the Bay Area who has collaborated with DJ QBert and Dan the Automator on custom hardware for performances. He's currently working on a series of inexpensive audio hacks with Jesse Dean Designs (who contributed the case/PCB files).

TIME REQUIRED:
15–30 Minutes

DIFFICULTY:
Easy

COST:
$15–$20

MATERIALS
» **STM32 "Blue Pill" microcontroller board** Arduino compatible
» **ST-LINK programmer**
» **Audio jack, 3.5mm, female**
» **Jumper wire** M/M if using breadboard, M/F if using header pins on STM32
» **Breadboard**

DRUM MACHINE:
» **Momentary tactile switches (4)** such as Alps #SKHHAKA010, or any other momentary switch — try arcade buttons!

SYNTHESLIDER:
» **Slide potentiometers, 45mm 10kΩ (4)** such as Digi-Key #BI PS45M-0MC2BR10K digikey.com

TOOLS
» **Computer** to run Arduino IDE
» **Soldering iron**

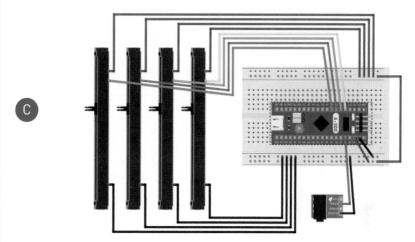
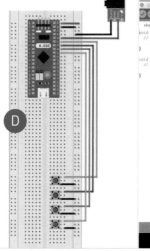
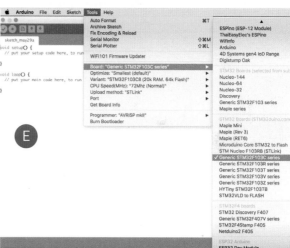

com/rogerclarkmelbourne/Arduino_STM32/wiki/Installation. Download the Mozzi sound synthesis library from GitHub: github.com/sensorium/Mozzi. Unzip and copy Mozzi to your Arduino libraries directory (**Mac:** *Documents/Arduino/Libraries*; **Windows:** *My Documents\Arduino\libraries*\) and launch the Arduino IDE.

Download the corresponding Arduino sketch for the Syntheslider (makezine.com/go/syntheslider) or Drum Machine (makezine.com/go/drummachine). To load it, specify your board in the Arduino IDE by selecting "Generic STM32F103C8" from the menu Tools → Board → STM32 Boards (Figure **E**). Set the Upload Method

to "STLink." Then connect your ST-LINK programmer via USB and upload the sketch to the Blue Pill. You should now be in business!

USING THE SYNTHESLIDER
The sliders are mapped in our sketch as follows:
» **Slider 1:** Oscillator Frequency
» **Slider 2:** FM (Frequency Modulation) Intensity
» **Slider 3:** Modulation Rate
» **Slider 4:** User-definable

Try adding new functions to the fourth slider — delay and other effects can have

some fascinating results. Going further, the Syntheslider zip folder also includes the files to let you laser-cut/rout your own case and make your own PCB for the sliders.

USING THE DRUM MACHINE
Press the buttons to fire off one-shot drum sounds and make a beat!

To load a drum machine sample, use Audacity to create a RAW 8-bit audio file, then use the *char2mozzi.py* script under *Mozzi/extras/python/* to convert your audio file to a sample array that can be loaded into the sketch by using a header file (read the full instructions in the *int8_t2mozzi.py* file).

Now add the synth and rock out. ❂

TIME REQUIRED:
24–48 Hours Printing +
2–4 Hours to Assemble

DIFFICULTY:
Easy

COST:
$70–$90

MATERIALS
For each speaker:

» **3D printed parts: egg enclosure (1), ring cover (1), grill (1) and flex feet (4)** Download the free 3D files at thingiverse.com/dr_frost_dk/designs. For each speaker, I use 700g–800g of PLA filament (about $20) plus a bit of flexible filament for the feet. You can modify the ring to fit your speaker, or just delete it.

» **Loudspeaker driver, 4Ω, coaxial** Use a 4" or 5" speaker of your choice. For the Sioux CS 100 PRO and 130 PRO drivers that I use, see my tips on the Thingiverse pages.

» **Speaker foam sealing tape, 500mm length** MDM-5 type or similar

» **Bullet connectors, 4mm, gold plated (2)** HobbyKing #AM1003A, hobbyking.com

» **Machine screws, 3.5mm, coarse thread (4)**

» **Speaker cable, 2×2.5mm², max 300mm length**

» **Foam, 10mm thick, 100mm×150mm**

» **Cyanoacrylate (CA) glue** aka super glue

TOOLS
» **3D printer (optional)** You can find a makerspace or send the files out to a printing service.

» **Soldering iron and solder**
» **Drill and 5mm bit**
» **Thread tap, 3mm**
» **Screwdriver**

HEINE NIELSEN is a 37-year-old tinkerer. He's always modding things in ways they weren't intended to be used, and never afraid to try something totally untested.

Written by Heine Nielsen

Egg Speakers

3D-print these awesome enclosures for audiophile sound

(A)

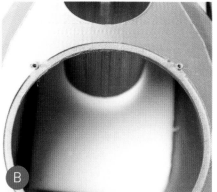

(B)

(C)

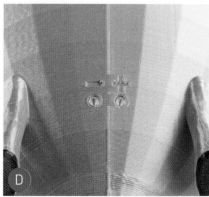

(D)

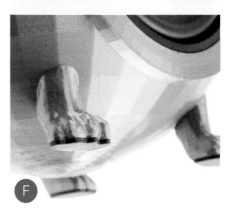

(E)

(F)

Hep Svadja

2. Drill out 5mm holes for the gold connectors in the back of the egg, for the speaker wire connection. (It's tricky to print these so they fit perfectly.)

3. Tap 3.5mm threads for the 4 speaker mounting holes, so the plastic won't separate when you screw in the speakers.

> **TIP:** If you can't find a tap to fit your selected screws, you can make a tap by grinding a slot in an extra screw. Check out woodgears.ca/thread_taps.

4. Put MDM-5 foam around the speaker hole (Figure A).

5. Put the 10mm foam behind the bass port. It should stay in place by itself (Figure B).

6. Solder 2 female gold bullet connectors to one end of the speaker cable.

7. Glue the gold connectors into the back of the egg (remember polarity), from the inside. This step is a bit fiddly, so do some dry runs (Figure C and D).

8. Connect your speaker of choice to the bare end of the speaker wire, then put on the ring, and screw it in (Figure E).

9. Finally, glue on the flex feet (Figure F).

Your first speaker is done; repeat these steps to build the second speaker.

USE IT

Connect your finished speakers to an amplifier, and enjoy! I'm very happy with the result — it sounds better than any other cabinet I've ever had in this size.

There are so many places to go from here. One thing I noticed: my original low-poly model creates some difficulties in having even wall infill-thickness. So I recently created a high-poly model too.

For my next build I'm printing these transparent, and adding WS2812B LEDs so they'll fill the room with sound and also with every kind of light and color pattern you can think of! ❷

After building speakers and amps for 20 years, I got into 3D printing last year and wanted to see if it was possible to make a good speaker enclosure this way. After looking at some 3D-printed parts my friend had made, and seeing how strong they could be, I was inspired to try.

The result: A pair of good-looking speakers that truly sound great, with no sharp corners to impede airflow inside the cabinet. The "egg" has always been one of the holy grails of cabinets in the hi-fi world, but they're hard to make in a conventional way. Now it's much easier with 3D printing.

I learned a lot about using infill settings to create an air gap between the inner and outer walls. This helps a lot — instead of just having solid plastic, the air gap dampens the pressure from inside, so the outer wall has less resonance. Material and wall thickness have the biggest effect on holding in the sound pressure, so you get more sound pressure in the listening room.

It's been a long process in getting the ratios "right" but I'm so amazed at the sound coming from such small speakers. I sell them for $170 a pair, but you can make your own pair for about $80. It's an easy build; only basic soldering skills are required.

BUILD A PAIR OF AUDIOPHILE EGG SPEAKERS

1. Take a minute to clean up your 3D prints so they look their best. This is a unique speaker cabinet you'll to want to show off.

Beauty and the Beats

Nothing less than a sound-reactive light show will do for Less Than Jake **Written by Matt Stultz**

MATT STULTZ
is *Make:*'s digital fabrication editor and the founder of 3DPPVD, Ocean State Maker Mill, and HackPGH.

At a Less Than Jake concert I noticed that Vinnie Fiorello's drums had LED lights inside. They would periodically change colors while the roadie set up the band's gear, but stayed one solid color during the performance. Less Than Jake have an energetic live show and these static lights just wouldn't do — I knew I could build them something better!

I needed to make something interactive but small enough to fit entirely into the drum itself. I also knew that there was no way any touring musician would want to pull out a laptop and reprogram an Arduino on stage, so I had to make it easy to make changes on the fly.

THE BRAINS

I chose the Huzzah32 not only for its tidy size, but because I also wanted to take advantage of its Wi-Fi and, potentially in later versions, Bluetooth interfaces. The lights are standard WS2812B LED strips with 60 lights per meter. To capture the drum hits, I used an accelerometer breakout board, the LIS3DH, which supports a tap mode (Figure Ⓐ).

I set up the Huzzah32 to be its own access point and server. The name of this access point ("DrumLights") and its WPA key ("HelloRockview") can be changed by altering this line in the code: `WiFi.softAP("DrumLights", "HelloRockview");`. When you connect, you'll automatically be taken to a webpage on your default browser (I had some issue with this on Android but it worked great on every other OS I tried). If it fails to launch, just point your browser to 192.168.4.1 to pull up the configuration page.

On the configuration page you can change three options: Flash, Brightness, and Sensitivity. Flash will make all the lights flash very brightly white for 0.1 second before changing color — a very dynamic effect, but it might be too much for some. Brightness allows you to change how bright the LEDs will be when normally on. Sensitivity determines how hard the drum

TIME REQUIRED:
1–2 Hours

DIFFICULTY:
Intermediate

COST:
$40–$60

MATERIALS
» **Huzzah32 microcontroller board**
  Adafruit #3405 adafruit.com
» **LIS3DH accelerometer breakout board** Adafruit #2809
» **WS2812B RGB LED strip, 60 LEDs per meter**
» **Stranded wire**
» **Solder**
» **Heat-shrink tubing**
» **Hot glue**
» **Zip tie**
» **USB micro cable**
» **USB power supply, 1.2A or more**
» **Case** Make your own, or 3D-print my files at thingiverse.com/thing:2958176.
» **3D printer filament (optional)**

TOOLS
» **Soldering iron**
» **Wire cutters / strippers**
» **Hot glue gun**
» **3D printer (optional)**

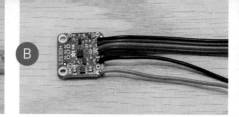

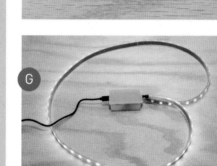

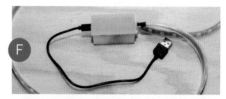

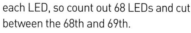

must be struck to trigger the LEDs; the farther to the right the slider is moved, the harder it is to trigger the color change.

THE INTERFACE

Set up your Huzzah32 in your Arduino environment. First, install the latest driver package from GitHub (github.com/espressif/arduino-esp32), following the instructions for your particular OS. Then you'll be able to select the "Adafruit ESP32 Feather" from the Board menu under Tools in the Arduino IDE.

You will also need to install the *Adafruit_LIS3DH* and *Adafruit_Sensor* libraries. These along with other documentation on the LIS3DH sensor can be found at learn.adafruit.com/adafruit-lis3dh-triple-axis-accelerometer-breakout/wiring-and-test.

To interface with the LEDs, you will need the Adafruit NeoPixel library, which can be installed from the library manager in your Arduino IDE. Make sure to restart Arduino after installing all of these packages.

THE BUILD

The WS2812B LEDs are best purchased in longer strips, but you only need about 44" for a 14" snare drum. There's a line between each LED, so count out 68 LEDs and cut between the 68th and 69th.

The soldering of this project is pretty straightforward (Figure **B**) with only one trick. You have two devices that need to be connected to ground, but only one ground pin on the Feather platform. I simply twisted my two ground wires together and pushed them through the hole and soldered them in place (Figure **C**). We are using the SPI interface on the LIS3DH board because at the time of writing, I2C was not working well on the ESP32 platform. This requires a few more wires but nothing that the system can't handle. See the wiring diagram (Figure **D**) for full wiring details.

I added a bit of support to the fragile connection where the wires connect to the LED strip. Using a spare piece of plastic (an SD card case), I cut a splint that I then hot-glued to the wires and the LED strip. With everything in its protective waterproofing tube, I capped each end of the LED strip with more hot glue to seal it. I used some heat-shrink tubing at the soldered end to keep it all together and looking nice (Figure **E**).

Once the board is wired up, download the code from github.com/MattStultz/DrumLights, open it in your Arduino IDE, and upload it to your board. Once complete, you should be able to tap on the sensor and see the lights change color. If you don't tap them for 30 seconds they will go into demo mode, changing color every 5 seconds.

Of course, you wouldn't want to just have these bare components dangling around in your drum so I designed a 3D-printable case (Figure **F**) for it all. The wired-up accelerometer is placed in the bottom of the case where it will be closest to the drum. The Huzzah32 is stuck to the inside of the lid with double-sided tape. There is a support for the LED strip on the side of the case that can be tightened down with a zip tie.

Finally, the Flash function pulls a lot of current — I found I needed a USB power supply able to output at least 1.2 amps.

ROCK IT

Now you're ready to shock the world with your snare drum light show (Figure **G**). I gave mine to Vinnie; he's testing it out! Attach the parts inside your drum however you like. I think good quality velcro is the best bet but hey, it's your instrument, make music and have fun with it. ◗

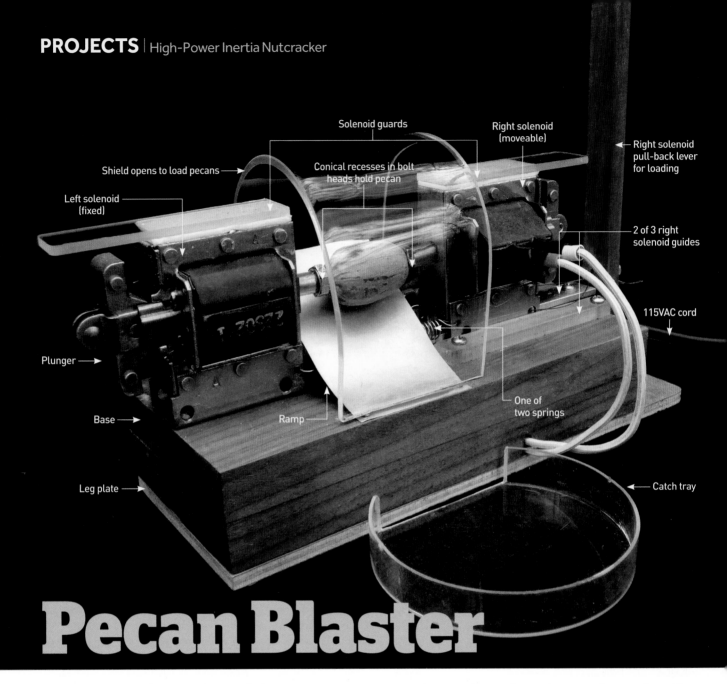

Solenoid guards

Right solenoid
(moveable)

Right solenoid
pull-back lever
for loading

Shield opens to load pecans →

Conical recesses in bolt
heads hold pecan

Left solenoid
(fixed)

2 of 3 right
solenoid guides

115VAC cord

Plunger →

Base →

Ramp

One of
two springs

Leg plate →

Catch tray

Pecan Blaster

Make a twin-solenoid power nutcracker that bashes the shells and saves the meats Written by Larry Cotton

LARRY COTTON has finally given up on doing anything earthshaking. He loves electronics, music and instruments, computers, birds, his dog, and wife — not necessarily in that order.

TIME REQUIRED:
A Weekend

DIFFICULTY:
Intermediate

COST:
$30–$50

So your Aunt Harriet brought you a big sack of pecans from her farm — awesome! Only one small catch: *you* get to shell them. No way are we going to do this manually with a "pliers" style nutcracker, right?

One alternative is to buy the legendary Texas Native nutcracker from Amazon or eBay for about $50. While this rubber band-powered relic from about 40 years ago does work, ask anyone who's cracked more than a dozen pecans with it how long it took for their left thumb to heal.

There are several semi-automatic,

one-pecan-at-a-time crackers on the market. All work using the same principle as the Texas Native: by bashing the pecan from one end, causing a weight at the other end to move a bit, with considerable resisting inertia. They can be bought for a paltry $200–$400 from vendors like Red Hill General Store. And you still have considerable follow-up picking to do.

The Pecan Blaster uses the same basic principle, but with two heavy-duty 115-volt appliance solenoids facing each other. Just load a pecan between the solenoids and close the clear protective shield to fire

MATERIALS

- » **Heavy-duty solenoids, 120VAC (2)** Dormeyer #7467S, from Electronics Goldmine (goldmine-elec-products.com), item #G20264B
- » **Wood, about 2"×4"×8"** preferably walnut or pecan, but pine is fine
- » **Wood scraps, about 6"×¾"×¼" (1) and 1"×⅝"×½" (1)**
- » **Plywood scrap, ¼" thick, about 2"×4"×9"**
- » **Acrylic sheet, ¼" thick, ½"×6" strips (2)** for side rails
- » **Acrylic sheet, ⅛" thick: ⅝"×¾" (2), 2¾"×9" (1) and ¾"×3¼" (2)** for rail spacers, shield, and solenoid guards
- » **Aluminum strip, ⅛"×⅝"×12"** for center rail
- » **Aluminum strip or extrusion, 1⁄16"×¼"×1¼"** for switch actuator
- » **Clear plastic salad box** for sliding surface
- » **Bolts, ¼"-20 or ¼"-28 (2)** with at least 1⅛" of smooth, unthreaded length
- » **Plastic ball-point pen housing** Paper Mate or similar
- » **Tension springs, 5⁄16"×1½"×0.023" (2)** Hillman #543470, available at Lowes
- » **Hinges, brass or steel: 1"×1" (1) and ¾"×¾" (1)**
- » **Rubber feet (4) or self-adhesive foam,** from Michaels, Hobby Lobby, etc.
- » **Micro switch, standard SPST, 115VAC** Omron, Honeywell, or Chinese knock-off
- » **Wood screws, #4×⅝" (2)** for micro switch
- » **Aluminum soda can** for the ramp
- » **Tiny nails (2)** for ramp
- » **Machine screws and nuts: #6-32×½"(3) and #8-32×1¾" (2)**
- » **Toothed washers, #8 (2)**
- » **Assorted sheet metal and/or wood screws**
- » **Wire terminals, female (4)** to fit your solenoids
- » **Power cord line, 115V** Reuse one from a dead appliance.
- » **Heat-shrink tubing**
- » **Electrical tape and double-sided tape**
- » **Soldering iron and solder**
- » **Nylon braided fishing line** or wire, to connect lever to movable solenoid
- » **Catch tray** of your choice

TOOLS

- » **Hammer**
- » **Pliers**
- » **Wire cutters**
- » **Sheet metal shears or scissors**
- » **Screwdrivers**
- » **Drill with drill bits and driver bits, assorted**
- » **Drill press**
- » **Steel-cutting countersink, ½" or ¾"**
- » **Sander or sandpaper, 120- and 320-grit**
- » **Handsaw or jigsaw** with wood and aluminum cutting blades
- » **Band saw (optional)** with wood blade
- » **Hacksaw and file**
- » **Hot glue gun**
- » **High-speed rotary tool (optional)** e.g. Dremel, with ¼" straight cutter
- » **Heat gun**
- » **X-Acto knife** with new No. 11 blade
- » **Vise** at least 4" capacity
- » **Rolling pin or PVC pipe, 2¾"–3" dia.** to bend shield around
- » **Center punch**
- » **Ruler, straightedges, circle templates**
- » **Sharp pencils with erasers**
- » **Wood finish (optional)** finishing oil, wax, Deft clear lacquer, etc. as desired

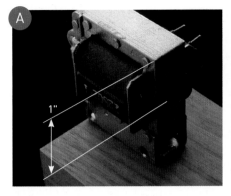

both solenoids simultaneously. The pecans don't stand a chance when bashed from both ends. The left solenoid is fixed, but the right one slides to accommodate various size nuts* and kicks back somewhat, using the inertial weight principle of the old-style crackers. Both halves of the nut usually survive unscathed and separated from most of the shell; the rest of the time I get quarters and smaller pieces, which are fine for most baking or snacking purposes. Most cracks need no further shelling.

Here's how you can build one like mine, but feel free to use alternate materials and slightly different measurements than what I chose.

> *Little-known fact: Pecans aren't true nuts; they're drupes. makezine.com/go/nuts-and-drupes

BUILD THE BASHERS

1. Buy your solenoids from Electronic Goldmine, item #G20264B, the cheapest I could find that would do the job. Two of these babies cost under $10. EG's got a $10 minimum, so look around for something else to add to your order.

Or, if you really want them in two days, Amazon has them for around $25 each (plus shipping), item #B01MQ4H6OP.

2. Make a base from an 8" length of 2×4. I happened to have some walnut, and of course it would be slick to use pecan! (Ask Aunt Harriet if you can cut a limb off one of her trees.) With a band saw or circular saw, cut a wiring slot about ⅛" wide across the bottom, about 2¾" from one end.

> **WARNING:** Don't energize the solenoids until all connections to them are insulated!

3. Prepare the solenoids. They're designed to pull, but they must push. So carefully drill, with a drill press, a ⅛" hole in the center of each solenoid in the plain side. The hole is 1" down from the top (Figure A), all the way through the approximately ¼" wall. Use a center punch for accuracy! Each hole must be in line with its respective internal plunger. Carefully widen the hole to ¼" using several bits from ⅛" to ¼".

After drilling, carefully blow, brush, and/or wash out with solvent (such as paint thinner; not water) any chips that fell into the solenoid. A magnetized screwdriver tip can help as well. Slide a ¼"-20 or ¼"-28 bolt back and forth through the solenoids' holes until they offer no resistance.

> **IMPORTANT:** The plungers must be able to easily push the bolts when the solenoids are energized, with no resistance.

While you're at it, also drill two 3⁄32" holes in the end of each solenoid to hold the ends of springs as shown in Step 8.

4. Mount one solenoid on the base's top, at the opposite end from the wire slot. It will be on your left as you use the Blaster and should be on the centerline, flush with the end of the 2×4. Keep it parallel to the long sides, with the solenoid's plunger hanging off the end of the base (Figure B). Mount it by drilling slightly oversize holes through the base to match the hole pattern in the solenoid. I used two 1¾" 8-32 machine screws, nuts, and toothed washers; tighten that sucker down.

5. Mount the other solenoid at the right end. It will slide back and forth, from about 2" away from the left solenoid to about 3½" away. Its plunger must be to the right. This solenoid must be guided to slide smoothly

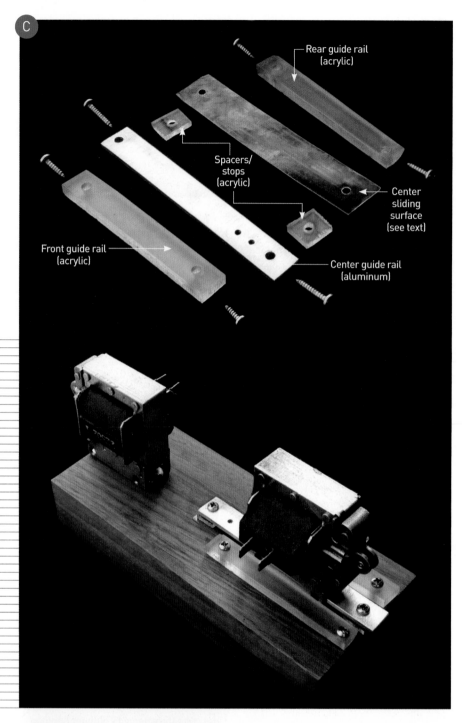

C

Rear guide rail
(acrylic)

Spacers/
stops
(acrylic)

Center
sliding
surface
(see text)

Front guide rail
(acrylic)

Center guide rail
(aluminum)

D

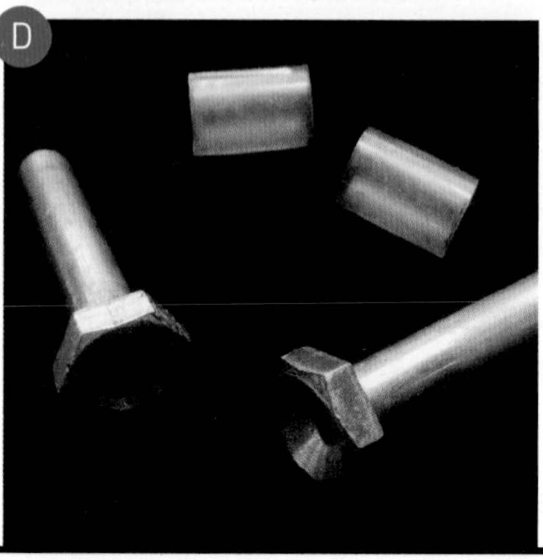

E

¼"-20 or
¼"-28 bolt

⁷⁄₁₆" sleeve

Conical recess

⁵⁄₁₆"–
³⁄₈"

1⅛" unthreaded

⅛"
pilot
hole

and accurately with minimum friction. I made 3 rails: two side rails from ¼" acrylic (such as Plexiglas) and a center rail from ⅛"×⅝" aluminum strip (Figure **C**).

The center rail's ends must be elevated ⅛" to prevent binding. I made two ⅛"×¾"×¾" acrylic spacers, which double as limit stops. Mount them so that the right solenoid moves smoothly back and forth without excessive play, and stops about 2" from the left solenoid at their closest positions. Travel distance will be about 1½". Later, two springs will connect the two solenoids.

IMPORTANT: The guide rails must restrain the solenoid from any side-to-side and up-and-down movement, while allowing free sliding. Tinker with this until you get it right. Drilling the strips' mounting holes slightly oversize allows for a bit of adjustment. I also glued a thin sliding surface (made from a rigid clear plastic salad box) to the base. You might also try a healthy spray of Teflon lubricant, such as Amazon #B00D3ZKVAS.

6. I made the pecan holders/bashers from ¼"-20 bolts (¼"-28 is OK, too). Use 2 bolts with at least 1⅛" of smooth, unthreaded length. Cut them with a hacksaw, and file the ends smooth so they'll easily go into the solenoids.

The heads need a conical recess to properly locate the pecans. With a drill press and vise, clamp each bolt and drill a ⅛" hole in the center of its head (use a center punch) about ⅛" deep. Switch to a 45° or 50° angle, ½" diameter (minimum) countersink bit, and countersink over the drilled holes until the diameter of the conical shape is ⁵⁄₁₆"–³⁄₈" at the tops of the bolts (Figures **D** and **E**). These cups are important to properly hold the pecans for bashing.

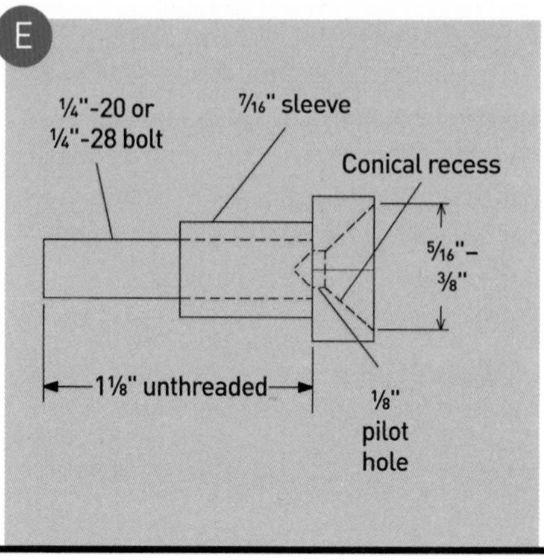

7. I made bolt-head spacers out of a ballpoint pen housing (think cheapest Paper Mate), about ⁷⁄₁₆" long. They should fit snugly over the bolts. If not, hot-glue them in place under the heads.

Poke the two bolts into the solenoid holes — they must slide in freely — and bring the solenoids as close as they can go together (about 2", remember?). The bolt heads should now be in line with each other, about ¾"–⅞" apart at their closest. In addition, when both bolts are pushed in until their spacers hit the

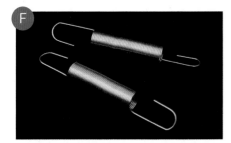

Drill ³⁄₃₂" both sides

Drill ³⁄₃₂" both sides

Left solenoid

Right solenoid

solenoids' laminations, the gaps between the outsides of their laminations to their plungers must be around ¼". This is the distance each solenoid would/could/will push its plunger (depending on the toughness of the pecan). If not, reposition the right solenoid or adjust lengths of bolts and/or spacers as necessary.

8. Make and install the springs. Actually, *alter* is the best word. Buy a pair of Hillman ⁵⁄₁₆"×1½"×0.023" wire diameter springs. Lowes stocks them. They're too long for this task, but about half-length is just right. Snip both of them roughly in half, so the stack of coils measures about ⅝". Bend out one coil on each for a hook (Figures **F** and **G**).

Hook the springs into the ³⁄₃₂" holes you drilled in the sides of the solenoids (Figure **H**). The right solenoid will now be held inward. Check your dimensions again as in Step 7 above.

MAKE THE SHIELD

A shield is necessary to keep pecan bits from flying toward your face. When you close the shield, it will turn on a switch to energize the solenoids.

9. Make the shield from a 2¾"×9" strip of ⅛" clear acrylic. I used a 2⅜"-diameter rolling pin as a form to bend the strip around. A piece of acrylic pipe or other cylinder of that approximate diameter should work.

Clamp your form in a vise and begin slowly to heat the strip in the middle on both sides (Figure **I**). As the strip heats up, it will begin to sag (Figure **J**). Be sure to apply enough heat in a 4"-long area in the center for the strip to form into a U shape around your form (Figure **K**). Take your time; getting it too hot will cause the acrylic to distort and/or bubble. Once you bend it around the form, hold it there, with

its two ends roughly parallel, until it cools and rigidifies.

NOTE: Harbor Freight heat guns for $9–$15, such as item #96289, work great for bending plastic, shrinking heat-shrink tubing, etc.

10. Because shield dimensions will vary, fitting your shield to your Pecan Blaster is a trial-and-error process. The goal is to mount and hinge the shield so that it swings freely from wide open (to allow placing a pecan) to all the way closed, completely shielding the opening between solenoids. When opening or closing, the shield must not touch the solenoids. As the shield fully closes, it (with an attached switch actuator) will turn on a 115VAC micro switch which energizes the solenoids.

Using 6-32 machine screws and nuts, fasten a 1"-wide hinge to the back center of the shield. Temporarily mount, with only a drop of hot glue, a small hinge-supporting wood block (about 1"×⅝"×½" high) about ⅜" from the back of the base, centered between the two solenoids in their closest position. Using the hinge as a template, drill two small holes in the top of the block and use wood screws to fasten the other side of the hinge to the block (Figure **L**).

Swing the shield from back to front, enclosing the space between the solenoids. Adjust the block's size and location to make this work. Try to keep the shield as low as possible. You may have to remove (temporarily) the hinge and trim either or both ends of the shield to make it more compact. Safety is paramount here, so take your time and do it right. Do not permanently mount the block just yet!

11. Add the switch and its actuator. As you did with the small block, temporarily mount the micro switch with hot glue as shown. I made the switch actuator from a scrap of ¼"×¹⁄₁₆" aluminum bent in a lazy Z shape. Attach it near the bottom edge of

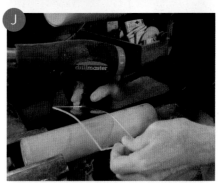

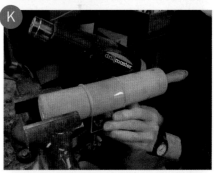

Larry Cotton, Jude Brown

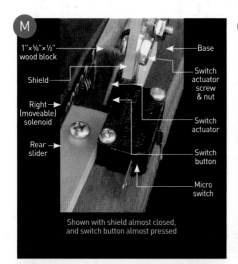

1"×⅝"×½" wood block — Base

Shield — Switch actuator screw & nut

Right (moveable) solenoid — Switch actuator

Rear slider — Switch button

Micro switch

Shown with shield almost closed, and switch button almost pressed

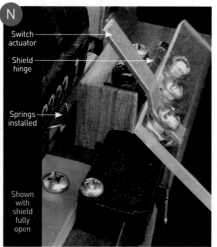

Switch actuator

Shield hinge

Springs installed

Shown with shield fully open

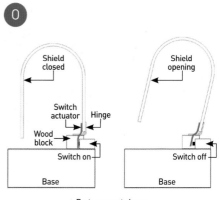

Shield closed — Shield opening

Switch actuator — Hinge

Wood block

Switch on — Switch off

Base — Base

Fasteners not shown

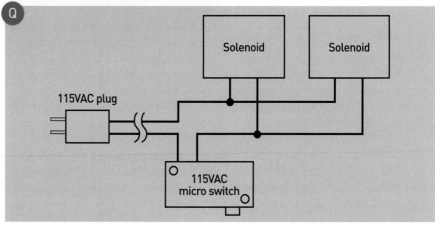

Solenoid — Solenoid

115VAC plug

115VAC micro switch

the shield with a 6-32 machine screw and nut so that when you close the shield, the actuator encounters the micro switch's button and presses it in fully. Adjust the actuator and/or micro switch to ensure they work perfectly every time (Figures M and N). When they do, permanently glue the block and switch actuator in their places and tighten all screws and nuts. Mount the micro switch with two #4×⅝" screws. (You may have to enlarge the holes to ⅛".) Finally, permanently glue or screw down the hinge-support block (Figure O).

FINAL ASSEMBLY

12. Make a channeling ramp from an aluminum soda or beer can, or any thin aluminum, about 2"×3½". Hold it temporarily at the front edge of the base, between the two solenoids. It'll probably rest on the closest spring, and that's OK! But make sure that the switch actuator misses the ramp when the shield opens and closes. Adjust the ramp or actuator length if there's any interference. Then, with two tiny nails, secure the ramp to the base (Figure P).

13. Make guards for the solenoids from ⅛" acrylic, ¾"×3¼". Clean the top surfaces of the solenoids, then use double-sided tape to attach the guards as shown on page 44.

14. Wire it! I followed a very basic schematic (Figure Q) that shows power switched to both solenoids simultaneously. Use female wire terminals that fit the male terminals on the solenoids. Use shrink tubing and plenty of electrical tape to insulate any exposed electrical parts. Route the wires from the right solenoid to the back through the slot in

the bottom of the base and secure all with wire hold-downs or staples (Figure R).

15. Make the leg plate from ¼" plywood, 1" longer than the base, then screw it to the bottom of the base. Ensure the crossing wires are in the slot. Attach 4 or 5 small rubber legs (Figure S).

16. Make a lever to pull the right solenoid to the right for loading pecans, from a small piece of walnut or other hardwood about 6" long. Drill a ³⁄₃₂" hole 1⅞" from one end. From that same end, attach it to the leg plate with another hinge like the one used on the acrylic shield.

Thread braided nylon fishing line or thin wire through two holes in the right solenoid, and over to the small hole in the lever. Use a screw to keep the string or wire from slipping. Leave a bit of slack in the line or wire (Figure T).

Test-pull the top of the lever to the right to open the gap between bolt heads. The solenoid must slide easily with no bind or excessive play.

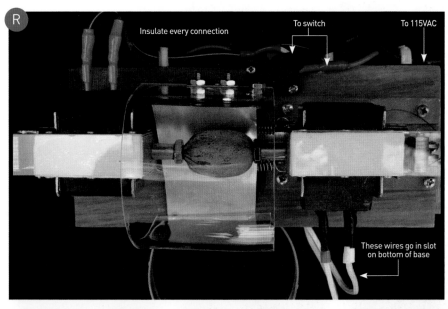

Insulate every connection
To switch
To 115VAC

These wires go in slot on bottom of base

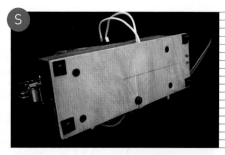

17. Make a catch tray. I used an old clock face and sawed and sanded a straight notch on one side about 2" long.

LET'S GET NUTS

Before you blast, test your Pecan Blaster by flipping the clear shield back, then plugging in the solenoids. *They must not fire.* If they do, unplug and check your switch and your wiring!

USE IT

When all is well, follow these steps:
a. Open the shield.
b. Use the hardwood lever to open or close the gap between the bolt heads.
c. Load a pecan and twist it a bit to ensure it's held securely by the bolts.
d. Important and at first unintuitive: Pull both plungers outward as far as they will go.
e. Close the shield to blast the pecan apart!

TROUBLESHOOTING

If the pecans don't crack on the first hit (they usually do), try again. Make sure you pull the plungers all the way out before blasting; it's easy to forget.

» **Sliding solenoid.** The right solenoid may not be moving freely, or may be moving too freely (side-to-side or up). Try pulling the lever back and forth a few times. The right solenoid must slide freely, directly opposite the fixed one. Just after the solenoids are fired, the right one should bounce a bit to the right in a straight line; it must not kick up or sideways.

» **Sticky bolts.** Also ensure the bolts slide smoothly in the holes you drilled in the solenoids.

» **Check dimensions.** Three dimensions are important:

1. The minimum distance between solenoids should be about 2" (Figure **U**).
2. The (empty) distance between bolt heads should be ¾"–⅞".
3. The gaps between the outsides of the laminations and their plungers must be around ¼" before you pull them out (Figure **V**).

Happy blasting! ⊘

¾"–⅞"

~2"

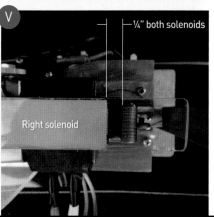

¼" both solenoids

Right solenoid

Larry Cotton

Written and photographed by Greg Treseder

GREG TRESEDER is a former aerospace engineer and founder of Fab Forge Five, a family owned digital fabrication company. He believes making is a fundamental part of human culture, and a great way to have fun.

Via Libris
How I built a mysterious GPS treasure book that guides you to a secret location — then out pops a gift!

Reverse Geocache Book

The Via Libris is a handmade, paperback-sized "GPS treasure book." It leads its user to a predetermined location, using a special compass dial and "cold," "warm," or "hot" clues. When the user reaches the treasure location, a gift card pops out of a hidden slot in the book. Surprise!

A new location can be programmed anytime with the push of a button. Each book is one-of-a-kind, beautiful, and reusable. I use mine to lead friends and family to the store or restaurant that the gift card is for. It's great fun. So I decided to turn Via Libris into a kit that others can build. I learned a lot along the way.

THAT OLD MECHANICAL FEEL

The Via Libris uses an Arduino Pro Mini microcontroller, a magnetometer, accelerometer, GPS module, sub-micro servo, and a custom-made continuous-rotation gearmotor servo (for the compass pointer). The project was inspired by Kenton Harris' reverse geocaching box published by Adafruit in 2012. I really liked the idea of leading someone to a location to find a surprise, which is in the box all along.

NOTE: *Make:* published Mikal Hart's original "Reverse Geocache Puzzle Box" in 2011. —*Editors*

The Via Libris was designed to be a bit old-fashioned. I love things that move; I like that good old mechanical feel. So I made a conscious decision to implement many features in hardware, using servos that rotate indicator wheels. An LCD display or LED array would have been much simpler, but I really enjoy building mechanical projects with Arduino.

Early on, I decided I only wanted two motors in my product, to save space and minimize magnets in a design that relied upon a sensitive electronic compass to find magnetic bearing. But this would challenge me to create three types of movement with only two servos — the compass wheel, the clue wheel, and the gift card dispenser.

DESIGNING A PCB

My prototype used electronic "modules": the Arduino Pro Mini, Adafruit Ultimate GPS, a servo controller, and an electronic compass. Rather than rebuild all those products by soldering tiny SMT components on one board, I used EAGLE software to design a master PCB onto which I could easily solder the various modules (Figure A).

Before sending my PCB design to a fabricator, I built my own board to test the system. It was crude, but it worked! I learned that a big ground plane under the GPS module is important — it isolates the weak GPS signals from the electrical noise that an Arduino (and servos) produce.

As soon as I received my PCBs I saw I'd made an error. I had to rework a few paths on every board. Triple-check before you order!

3D DESIGN IS KEY

Remember my decision to use two servos for three different motions? This created a complex design challenge. I needed the clue wheel servo to also trigger a dispenser to "pop" the gift card from a hidden slot in the book, when the clue wheel rotated to the "you've arrived" position.

At first I decided not to do a 3D design, and built prototypes instead. I tried five different dispensers. Each would either work well but be difficult to make, or would be easy to make and work intermittently.

When version five was a failure, I threw up my hands and decided to invest my time in creating a good 3D CAD model. I spent a few hours creating a complete assembly of my design, minus a dispenser, in Autodesk Fusion 360 (Figure B). This allowed me to visualize the space constraints and try different dispensers in my 3D model without building anything. If it didn't work, I simply clicked Undo. I finally settled on a reliable solution, using three 3D-printed parts. In total, I spent about 10 hours on this design process. Had I done it at the beginning, I would have saved months of aggravation.

CODE TWEAKING

The Arduino code had to handle all the I/O for a variety of external devices. The GPS module uses serial communication. The electronic compass uses I2C. The servos use digital pins, and my custom servo also needed an analog pin to measure feedback. There's also a small vibrator to provide user feedback, which uses another digital pin.

The Arduino uses its serial port for programming, but the GPS module also uses that serial port, so it interferes with uploading a program. I tried shifting the GPS to a couple of digital pins using the SoftwareSerial Arduino library to emulate the hardware serial interface. Then I found that soft serial and servos interfere with each other's timing. So, I added some simple jumpers to my PCB to disconnect the GPS for programming — a hardware change to solve a software shortcoming.

DOCUMENT EVERYTHING

During my build process, I documented every step with photos and notes, including revision histories, all parts, and any special tool setups. I therefore created an assembly and manufacturing guide on the fly. I highly recommend this. You'll need it when you ramp up your production, even if you're building only 10 items for family and friends. Also, add lots of comments to your code.

FRIENDS + FAMILY = GUINEA PIGS

Once I had working prototypes, I packaged them into a nice "book box" from a craft store (Figure C). I chose a few adventurous friends and family members, and gave each a Via Libris pre-programmed for a merchant location near them, with a pre-loaded gift card. I wanted to see if they could just turn on the box and follow it, and I wanted them to be surprised. It worked!

I received some important feedback. My box doesn't tell the user how far they are from the "treasure" location. Car? Bike? Walk? So I added icons that represent a car, or footsteps. I also discovered a subtle error in the trigonometry function that calculates bearing. It wasn't apparent around my home latitude/longitude, but was a definite problem at my mom's house!

It's been a fun and educational process. I wrote a prototyping guide with lessons learned; I hope it inspires you to tackle a project you've been dreaming of. ●

[+] Read the full version of this article at makezine.com/via-libris and get my Prototyping Guide at fabforgefive.com/treasure-book.

TIME REQUIRED:
2–3 Hours

DIFFICULTY:
Intermediate

COST:
$90–$100

MATERIALS

Via Libris GPS Treasure Book Kit
$90 from fabforgefive.com/store. Key components include:
» **Arduino Pro Mini microcontroller board**
» **GPS module pre-soldered to custom printed circuit board (PCB)**
» **LSM303 eCompass module** with 3D accelerometer and magnetometer
» **Micro servomotor**
» **Continuous-rotation servo gearmotor**
» **Rotary sensor**
» **Piezoelectric vibrating disc**
» **Pushbutton switch**
» **Slide switch**
» **Laser-cut book box**
» **3D-printed mechanisms and brackets**
» **Laser-cut acrylic windows and wheels**
» **Various screws and hardware**

TIME REQUIRED:
1–3 Hours

DIFFICULTY:
Easy

COST:
$80–$150

MATERIALS
» **Corrugated cardboard**
» **Heavy duty plastic trash bag** Transparent bags are easier to work with.
» **Servomotor, 9g micro size**
» **Electronic speed controller (ESC), approx. 10A**
» **Brushless motor, 1,000kV–2,000kV**
» **Propeller, size 5×4.5**
» **R/C transmitter and receiver, 2 channel minimum**
» **Double-sided tape, strong**
» **Kids' toy beach tennis rackets (2)** or similar. I find these at the dollar store.
» **Strawbees connectors (34)** Find a local retailer at strawbees.com/resellers.
» **Drinking straws (3)** Reuse old straws, or get Strawbees recyclable straws.
» **Rubber bands** or zip ties
» **Scotch tape**
» **3D-printed parts (optional)** for the motor pod. Download the 3D files from makezine. com/go/cardboard-hovercraft. Or make your own from cardboard or 1mm–2mm plastic.
OR PICK UP A KIT like those pictured above at strawbees.com/store

TOOLS
» **Paper templates** Download them from makezine.com/go/cardboard-hovercraft.
» **Pencil**
» **Box cutter**
» **Scissors**
» **Phillips screwdriver**
» **Metal ruler**
» **Saw**
» **Hot glue gun or epoxy (optional)** if you're making the DIY cardboard motor pod

DIY R/C Hovercraft

Made from cardboard and trash bags, these things rip! Make a bunch and race your friends **Written by Erik Thorstensson**

ERIK THORSTENSSON is an engineer, entrepreneur, and co-founder of Strawbees, an award-winning prototyping toy for all ages.

It all started when I crashed a DIY quadcopter and wanted to reuse the parts for something fun. This was on a weekend when I had a slight fever, so I had some spare time on my hands … and I had an old dream of building a hovercraft … as makers, you know how it goes!

So I put together some cardboard, trash bags, and electronics, and it just worked. Since then I've been refining the design, trying out different things and having lots of fun building hovercraft as big as 1 meter square. I've also had help from some awesome Norwegian makers from Makeadrone (makeadrone.net), who immediately started holding workshops building some of my early designs and nudged my company, Strawbees, in the right direction with improvements. Finally, we released it to the market as an educational experiment kit.

The entire idea of our cardboard hovercraft construction is having an easy platform for experimentation. We use a snap-on, snap-off motor system and low-cost materials so you can easily make fast, iterative design changes to the base, skirt, and steering fin. It's a ton of fun, and

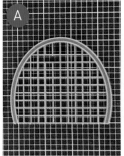

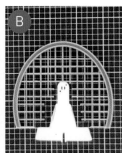

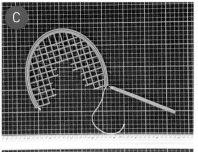

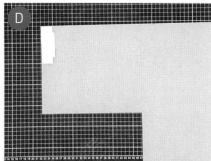

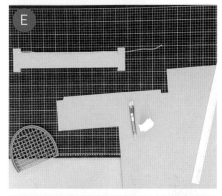

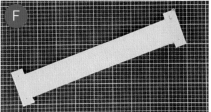

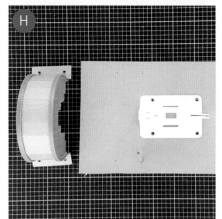

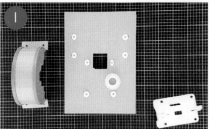

hopefully it will help you learn some great product development skills.

And these babies are fast! They hover across land and sea at over 25mph (~40km/h). Create your own spectacular designs, and race them with your friends.

1. MAKE THE PROPELLER GUARD

You'll start by building the items that will mount on top of the base, so you'll know where to put your holes in the base. First, the propeller guard.

Find cheap toy "beach tennis" rackets made in durable plastic. Many of these are molded with a slot that's perfect for cardboard; if not, you can just tape or glue the cardboard to the edge. A cardboard guard, in combination with the plastic grid, will stop things from hitting your propeller (and the other way around too of course). Make sure there's room for your propeller to spin freely, then mark your cut and saw off the rackets at this point (Figure A).

On one of the rackets, you need to make room for the motor pod. Align the pod and mark around it with a pencil, then cut on your line (Figure B).

Place a length of string in the slot in the edge of the racket, and mark it to measure how long the cardboard for the guard should be (Figure C). You can also use a flexible measuring tape.

Cut out the end template and place it on a piece of corrugated cardboard, parallel to the fluting in the cardboard (Figure D). This makes it easy to slot it into the racket later. Mark the first end's contour and holes using the template. Then measure the distance (that you marked on the string) to the other side and mark the opposite end using the same template.

Finish the contour by cutting the ends along your marks and then cut the long edges between them (Figure E). Round the corners if you want it cleaner looking; we chose not to in this tutorial.

Punch the four holes out with a pencil (Figure F).

Slot the cardboard in between the two rackets (Figure G). Secure with rubber bands or zip ties. Your propeller guard is now done.

2. PREPARE THE BASE

Use the base template or measure a 25cm × 34cm square. For strength, make sure the fluting in the cardboard runs front-to-back on your hovercraft; otherwise it folds too easily in a crash. Cut the square out using the metal ruler.

Cut the 5cm × 5cm hole for the motor pod according to the template. Take note of the offset from the center; this position is important for the balance of the hovercraft. Depending on the weight of your electronics and propeller guard, you can try out different offsets to perfect your design. If your hovercraft is tail-heavy, move the assembly forward on the next base you make, and if it's front-heavy, move it backward. You can also tape weights to your hovercraft to balance it before your build your next one.

Place your motor mount (see Steps 3 and 4) on top of the hole and mark the 4 holes for the motor pod. Do the same for the propeller guard. Make sure the propeller guard leaves enough room for the motor and propeller. Punch the holes with a pencil (Figure H).

Choose which side is going to be up, then thread single-legged Strawbees connectors up through all the holes (Figure I). Fold the round head of the Strawbees flat against the base bottom and secure with tape. You need 8 Strawbees per base; you can reuse them after this one is trashed.

3. 3D-PRINTED MOTOR POD

If you have access to a 3D printer, follow this step; if not, go to Step 4.

The 3D-printed motor pod comes in two

Erik Thorstensson

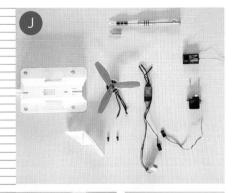

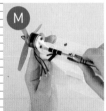

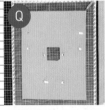

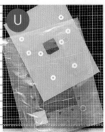

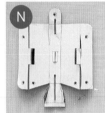

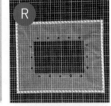

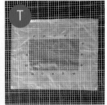

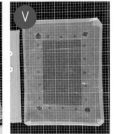

pieces: the servo goes in the *motor mount*, and the brushless motor goes in the *motor pod* (Figure J). Modify the mounting holes if necessary to fit your motors.

Mount the servo and then connect it to the corresponding channel on the receiver. Put the receiver and electronic speed control (ESC) into the motor pod and secure with double-sided tape. Make sure there's room for the battery (Figure K).

Thread the cables from the ESC through the hole in the motor pod one at a time.

Attach the motor mount to the pod, and secure with Strawbees (Figure L).

Mount the brushless motor using its mounting screws or another system depending on what kind of motor you find (Figure M). Skip to Step 5.

4. CARDBOARD MOTOR POD

If you don't have access to a 3D printer, you can easily make this cardboard version.

Print, trace, and cut out the motor pod template in cardboard. Gently cut and pre-fold all the creases (Figure N).

Glue the pod together where indicated on the template (Figure O). Make sure to strengthen the mounting surface for the motor and secure it with epoxy or a glue gun.

For your brushless motor, cut out a cardboard mounting plate and add some plastic or wood to make it stronger. We used a milled wooden piece for this tutorial. Improvise to suit your motor setup. Just make it strong enough to hold your motor and make it easy to zip-tie into place.

Attach the motor to your mounting plate. We used screws but this depends on your setup. Some motors have a motor mount included. Then zip-tie the motor mount in place on the pod (Figure P).

5. MAKE THE SKIRT

Cut off the sealed edge of the trash bag and one of the folded edges, and open the bag so you're working with a single layer of plastic.

Place the hovercraft base on top of the plastic. Trace an offset outline of your base, about 4cm–5cm to the outer edge. This can be varied for different lift heights above the ground, varying the hovercraft's ability to cross obstructions but also its stability. Experiment as you see fit.

Place double-sided tape all along the outline of the edge. Cut the plastic at the outer edge of the tape (Figure Q).

Cut out an offset square in the center of the sheet, about 8cm from the outer edge. This also defines the lift height of the hovercraft.

Now you'll make the holes that let air out under the hovercraft. All around the center hole, mark the air holes about 1.5cm away from the edge of the inner square and 5cm between neighboring holes. Cut out each air hole about 1cm in diameter (Figure R). You can do many experiments with size and placement of these holes.

Peel the protective backing from the double-sided tape and then lay an additional plastic sheet on top of the bottom plastic sheet. Align it right along two edges to reduce waste (Figure S). Have a friend help out so it doesn't get stuck in the wrong position. Then press down with your hands and smooth out any creases.

Cut off the excess plastic along the outside of the double-sided tape (Figure T).

Mark 4 large holes on the top sheet approximately 3cm in diameter, inside the cardboard perimeter but outside the small holes (Figure U). Cut them out, making sure only to cut through the top layer of plastic. This is where the air from the motor pod is let into the skirt. The small holes are in the bottom sheet.

Turn the skirt inside out so the taped crease is on the inside of the skirt (Figure V).

Put double-sided tape along the outline of the bottom of the base. Center the cardboard base over the skirt and then press it down (Figure W).

6. MAKE THE STEERING FIN

Trace any of the fin templates on cardboard. Gently score the crease and bend it back and forth a couple of times so the fin can move easily.

Use a pencil to punch the holes for fixing the fin to the motor pod and the linkage. We cut the short fin model for a faster-turning, but harder-to-control hovercraft. The fin is really easy to experiment with, as it is only connected to the pod linkage with two Strawbees.

Use 4 Strawbees to make two cardboard connectors for the linkage, as shown in Figure X.

Attach the Strawbees cardboard connectors by pushing the leg side through and locking the legs with two more Strawbees on the other side of the cardboard. Squeeze the locking Strawbees

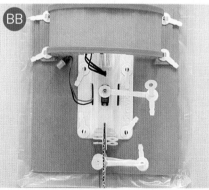

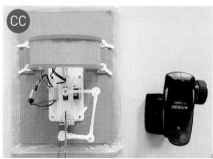

up against the cardboard so they sit tight and secure (Figure Y).

7. ASSEMBLY

Now you have all the pieces you need. It's time to assemble your hovercraft.

> **CAUTION:** The propeller moves very fast and could hurt you. Keep the battery disconnected until the hovercraft is ready to test, to prevent the motor from spinning.

Slide the motor pod onto the four connectors around the hole, then attach the propeller guard to its four connectors. Secure both with Strawbees from the top. Slide the Strawbees all the way down until everything sits tight against the cardboard base.

Put a Strawbee over the servo arm and secure it to the servo horn with a piece of drinking straw cut to about 5cm long (Figure Z).

Attach another straw to the top connector on the fin. Cut this one a bit shorter than the straw on the servo; we cut it to 3cm but you can experiment with the ratio to achieve the right amount of throw for your fin.

Then cut a straw to form a triangle from the bottom of the fin (Figure AA).

Put Strawbees in the three open ends of the straws. Then make linkage connections as shown (Figure BB). Leave the final linkage open until you power up the system.

Leave one motor cable disconnected from the ESC. Grab your R/C transmitter and make sure all trims are neutral, then connect the battery to power up the hovercraft. Secure your servo horn at 90° from the motor pod, check that your movement is correct, and reverse if necessary. Then cut the final straw linkage so that the fin lines up perfectly straight with the hovercraft (Figure CC).

Unplug the battery, then connect the cables to the motor. Your hovercraft is done.

USE IT

» Setup and Racing

Now follow the instructions that came with your speed controller to set up your R/C system. After that, you're all set to race!

Your hovercraft is cheap, fast, and surprisingly robust, able to survive multiple crashes at speeds up to 25mph (40km/h). And if you do break it, you can just upcycle more cardboard and trash bags to revive it.

It runs really well on flat surfaces, tarmac, and concrete.

» Upgrade for Water Use

With a few modifications your hovercraft runs really well on water too, but be warned that the cardboard base is no good for water. Instead, use thick packaging foam (Figure DD) as a great hovercraft base for water racing.

TIPS AND TUNING

Make sure the skirt inflates well. If it's too tight you can try stretching the bottom plastic a bit to create more room for the air to find the four corner holes.

Start by racing slow and learning how to control your hovercraft, then increase the speed. But always make sure to race in a safe space.

It's cheap and easy to modify the skirt, fin, linkages, and prop guard (air duct) length to tune your craft to your liking (Figure EE). I really hope you'll have a lot of fun racing and experimenting with your cardboard hovercraft! ◗

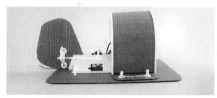

Erik Thorstensson

Building Furiosa's
Bionic Arm

Inspired by *Mad Max: Fury Road*, I built a real prosthetic arm for an amputee cosplayer

Written by Michelle Sleeper

E very once in a while, a character in film inspires you to make something you could only dream about.

Back in 2015, I was introduced to a cosplayer who wanted a truly unique prop made. Laura from Amplitude Cosplay is a left arm transradial amputee, meaning that she is missing her left arm below the elbow, since birth. She's not afraid of displaying her unique body with her costumes — we had originally connected because she wanted an exposed arm endoskeleton like from the film *Terminator*, and that was at the top of my list of dream projects.

Then the movie *Mad Max: Fury Road* came out, and with it the character of Imperator Furiosa, and she changed all our plans. Laura had authored an in-depth piece on her blog about representation of amputees and how Furiosa was masterfully written. Beyond that, her unique bionic arm

MICHELLE SLEEPER is an Atlanta-based maker, artist, and educator. She specializes in special FX and prop and costume making. You can see her work at msleeper.com and on Instagram at @overworlddesigns.

was completely awesome, and both Laura and I knew that we had to bring both it and her to life.

I have made plenty of prop and costume pieces before, but this was the first prosthetic I had ever made, and there were a lot of really exciting challenges. Unlike a space gun or video game armor, there wasn't a lot of wiggle room for fitting — the bionic arm had to fit her perfectly. So to start out we made a simple life cast of her arm. The mold was made using alginate, a cheap and easily found material for life casting, and from the mold we cast a duplicate out of plaster.

In order to design the 3D model accurately to fit Laura, we then used the plaster cast to create a 3D scan using photogrammetry software. Using the life cast was far more ideal than scanning Laura directly; having her sit perfectly still while we performed a scan would've been next to impossible. The plaster cast also doubled as a stand-in for her during the fabrication process — since I couldn't test-fit the prop on myself, the plaster cast was used to make sure everything fit in between Laura's visits to my shop.

Once the digital model was finalized and as close to "screen accurate" as we could get it, we sent it off to the 3D printers. A lot of the detail parts were cast from real-world items, such as the wrench tied to one of the forearm shafts, and other parts were laser-cut out of acrylic and heat-formed into their proper shape. Finishing up the parts from there was fairly standard post-processing procedure, and once everything was sanded smooth and painted, we then had to attach it to her.

The mechanical arm seen in the film is a mixture of practical and digital effects (there's actually only one shot in the whole film where the arm is fully practical!), but since we couldn't use movie magic in real life at a cosplay convention, we had to toe the line between accuracy (to the look of the prop in the film) and actually making it wearable. A few key scenes helped us understand how the film prop was worn; there was a series of leather straps running up the arm to her shoulder, and from the shoulder to a waist harness. The leather re-creation worked really well, and Laura, who doesn't normally wear a prosthetic, says it was comfortable.

On the day of the convention I delivered the prop to Laura and got to see the final costume all together. The finished product was astounding, and it was a lot of fun catching con-goers doing double takes as they tried to figure out how she was hiding her "real arm." Even when the truth was revealed, a lot of people still thought there was some sort of trick, but the reality was so much better — an empowering and real-world prosthetic for a special cosplayer. ◗

Life casting Laura's arm.

The hand, fresh off of the 3D printer.

Early stages of sanding, filling, and priming the prop arm.

The bionic arm finished and ready for battle in the Wasteland.

Close-up detail of the wrench, molded and cast from a real wrench!

Test-fitting the arm on Laura before we fabricated the leather harness.

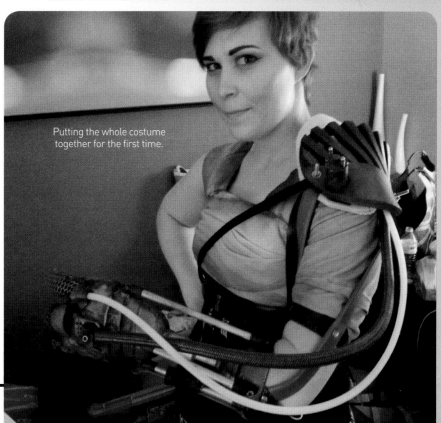

Putting the whole costume together for the first time.

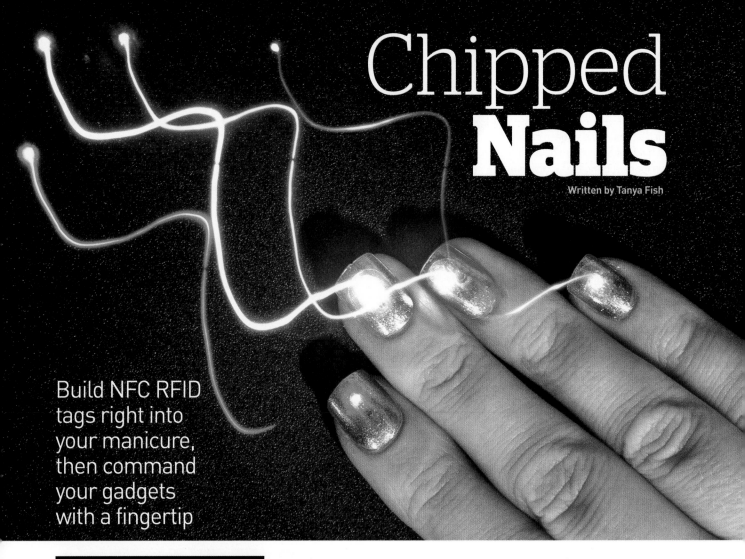

Chipped
Nails

Written by Tanya Fish

Build NFC RFID tags right into your manicure, then command your gadgets with a fingertip

TANYA FISH makes stuff because she can't help it. Formerly in U.K. schools teaching maths, physics, and shouting loudly at paperwork, she's now a crew member at Pimoroni, making learning materials for schools and workshops, and lasering all the things.

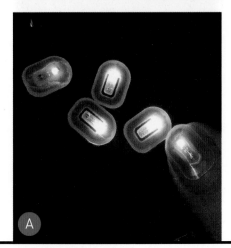

(A)

Nail art has seen some advances in the past decades, from the invention of cellulose-based polish 100 years ago, through thin plastic false nails, to the available-everywhere acrylic sculpted nails. But it was about time nails went techno.

On the security conference circuit, Baybe Doll (aka Emily Mitchell) had been getting her nail technician to embed small devices with readable data into her acrylic nails. However, the technology wasn't readily available to the masses, and when it was, it was big and chunky.

NFC (near-field communication) tags are the solution; they're tiny and they're powered by nearby magnetic fields so they don't need batteries. The first NFC tag I tried to put on my nail was an NXP Mifare Ultralight C NTAG213 that I bought from a supplier to retail stores (it's the thing that makes your shopping bag go beep when you leave if it hasn't been deactivated). It was

huge in comparison to the ones I use now.

I started searching online for tiny NFC data tags and LED lights, and tried out a variety before settling on the NXP Mifare Classic 1K, which is only 9mm across and can store a surprising amount of data.

The main problem was how to protect the tags and LEDs from handwashing and showers whilst on your nails. Painting over them works — but it's bumpy. Having them embedded in a sculpted nail by a nail technician is costly. This DIY method is a cheap and workable solution and, if you're careful, can be reused.

BUILD YOUR NFC/RFID NAILS

1. Glue your nail sticker — NFC chip or LED — to your natural nail (Figure Ⓐ), using the sticker adhesive (if it has one) or nail glue.

2. Glue a false nail over the top (Figure Ⓑ), and paint. Or get your nail technician to

Sandy Macdonald, Tanya Fish

TIME REQUIRED:
1 Hour

DIFFICULTY:
Easy

COST:
$15

MATERIALS
» **Set of 10 NFC LED nail stickers and/or NFC data tags** Pimoroni sells both for about $1 each (pimoroni.com), or you can find them cheaper online in big quantities.
» **Acrylic nails or stick-on false nails** You can make your own acrylic nails but you might prefer a professional nail technician.
» **Nail glue**
» **Nail polish**
» **Decorations**

OPTIONAL:
» **Acrylic liquid (ethyl methacrylate) and acrylic powder (poly ethyl methacrylate)** if you're making your own acrylic nails

TOOLS
» **Nail file**
» **Orange stick** or other wooden stick, e.g. BBQ skewer
» **High-speed rotary tool** e.g., a Dremel
» **Paintbrush**
» **Shot glass**

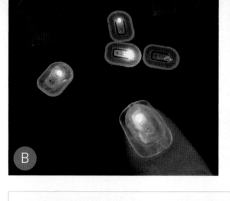

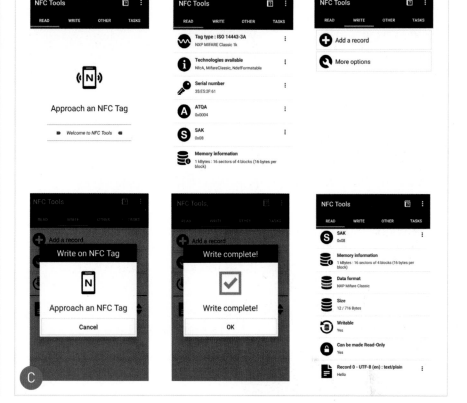

sculpt an acrylic nail over the top.

Or, DIY it — I used acrylic liquid and powder, which creates a moldable blob of plastic that you can smooth over the tag and file into shape before painting. Take care that you don't file any of the components — if you can see them sticking out, your acrylic isn't deep enough yet.

3. Program your nail sticker using NFC Tools (Figure C) or NFC Tasks (free apps from Google Play).

This won't work on an iPhone, but if your phone doesn't have NFC, Adafruit has created some Python libraries for the Raspberry Pi and BeagleBone, which you can use with their PN532 breakout board.

USE IT!
Use your data nails to trigger tasks on your phone or your homebrew NFC project (Figure D).

Or use the LED nails to show that your NFC card is being read when you hold it near a reader (Figure E).

NAILPAY?
I was intrigued by what else could be embedded in nails, so I soaked a credit card in acetone to find the chip inside. It was a very messy process, but eventually I got the chip out. Now I don't have to remember my wallet, although if I want to protect my data from accidental reads, I'm going to have to wear a thimble!

POWERED PCBS
I would like to have even more technology in my nails. I'm currently working on developing some flexible PCBs that attach to your nails and are powered by a matching bracelet. ●

CALEB KRAFT
is a senior editor
for *Make:*. He enjoys
learning new skills
just long enough
to fully understand
how talented true
craftspeople are.

Written by Caleb Kraft

Reverse Engineering Furniture

How I created my cheap CNC homage to a famous mid-century modern writing desk

A

Sometimes you see a picture of something and you think, "I must have that!" This is what I felt when I saw this random picture of a writing desk (Figure **A**), designed in the '50s by Helmut Magg. I thought the design was stunning and simple, and would fit right in, in my mid-century modern home.

There's just one problem. OK, two problems. This desk hasn't been produced in years, and the few that are circulating for sale cost thousands of dollars. I absolutely hate spending money, and I also know that

I'm hard on furniture (so are my cats). I don't even *want* that fancy thing in my house.

So what could I do? Reverse engineer the desk to build my own that pays homage to Helmut's design. I looked at the few pictures that there are available of his version and started drawing up my own plans.

WHAT I LOOKED AT
I studied the picture of Magg's desk. I knew that the **side profile** is really what I was after (Figure **B**). I really enjoyed the curve at the front edge, and the curve leading up to the

TIME REQUIRED:
A Weekend

DIFFICULTY:
Intermediate

COST:
$55–$75

MATERIALS
» **Appearance-grade plywood, ¾"** thick, 4'×8' sheet
» **Wood dowels**
» **Wood glue**
» **Wood finish** I used Polycrylic.

TOOLS
» **CNC router**
» **Computer**
» **Cutting files** Download my drawings from makershare.com/projects/mid-century-modern-writing-desk and modify them as you see fit.
» **Hand drill and sandpaper**
» **Dowel jig (optional)**

rear shelf (Figure C).

I looked at the **grain** in the wood to give me clues as to how it was constructed.

I looked for obvious **joinery** to see if there were any big "gotchas." Were there pocket holes or dovetails visible? Nope, nothing to learn there.

Ideally, I'd have the piece of furniture to rip apart and see how it was built. If I already had this desk though, this entire exercise would be pointless.

DECISIONS BEFORE DESIGNING
I'm not an experienced furniture builder so I knew I would have to design for my skills. While I do have a CNC router available, you could cut all these profiles with a jigsaw. They just wouldn't be as perfectly straight (if you're as clumsy as I am).

» **Legs** — Magg's desk legs have this fantastic flow of the grain from the bottom tip all the way to the top. It was obvious that his sidepiece was split in two, allowing for the grain to meet in the middle. I really wanted this effect (Figure D), even though I knew I could've just cut a single piece. Instead, I split mine like his, attaching the two halves using dowels (Figure E). (I happen to have a nice dowel jig that I made on my Tormach mill, inspired by a video by David Picciuto.)

» **Drawer** — The drawer on Magg's design is nestled in the front, and frankly, I didn't think I needed one, so I simplified the design to just remove the drawer altogether.

» **Desktop and shelf** — These are simple rectangles. I could have measured and cut them on the table saw, then attached them, again, with biscuits or dowels. This would have required careful measurement.

However, as I stated before, I've got a CNC machine! I took advantage of the ability to design in Autodesk Fusion360 allowing me to create *blind tenon joints* for all my

horizontal surfaces. Now, I admit I didn't even know what those were called before sitting down to write this, I kept just calling them "pocket holes" which happens to be an entirely different thing in the world of woodworking. A blind tenon is where you have a hole (mortise) in the wood that a tab (tenon) fits into. The *blind* part means it doesn't go all the way through, resulting in a hidden joint. Just like Magg's, my desk would have no visible joinery!

MY RESULTS
After double- and triple-checking my design, I bought a single sheet of ¾" maple plywood and ran the file on my router. To my surprise, everything went perfectly smoothly. A quick test fit showed that my blind tenons would fit together with only a quick sanding to clean things up.

D'OH!
The only problem I ran into was the fact that I completely forgot to add the mortises for the top shelf, and subsequently missed that obvious oversight on all my double-checkings. To remedy this, I just cut off those tenons (the tabs that go into the holes) and glued the shelf in place.

WHAT WOULD HELMUT THINK?
The desk now sits in my office, and gets used regularly. I quite enjoy the aesthetics and feel proud that I was able to pull it off. If I were to improve on this, I'd probably ruin the lines a bit by adding a stretcher across the back legs to help reduce wobble from the plywood construction.

I do find myself wondering how Helmut Magg would feel about all this. Honored? Amused? Irritated? A big part of the mid-century modern design philosophy was to design for the new tools of mass production; my home-hobby CNC version is like a new evolution of that philosophy. Unfortunately, Magg passed away in 2013, so I'll never know. ◐

See more of Caleb's mid-century desk (and his dowel jig) at makezine.com/2018/03/02/mid-century.

Caleb Kraft

TIME REQUIRED:
A Weekend

DIFFICULTY:
Intermediate

COST:
$100–$120

MATERIALS

» **Copper sheet, 20 gauge** (.0217 inches thick), 24"×24"
» **Lead-free plumbing solder, 1 large spool** I recommend a cadmium-free, food-safe solder such as Stay-Brite 8 or Bridgit.
» **Lead-free, water soluble plumbing flux** such as Bridgit
» **Soft copper tubing, ¼", 20' roll**
» **Silicone rubber stopper, #7 size**
» **High-temperture silicone sealant, 1 tube (optional)**
» **Brass machine screws, Phillips head, #8-32 × ¼" long (13)**
» **Brass nuts, #8-32 (13)**

TOOLS

» **Flat leg divider**
» **Metal scribing tool**
» **360° protractor**
» **Ruler**
» **Propane torch**
» **Flux brush**
» **Locking pliers, medium sized (2)** such as Vise-Grips
» **Aviation snips** For right-handers, I suggest offset, red-handled snips for cutting counterclockwise.
» **Small hammer**
» **File**
» **Sandpaper, medium grit**
» **Cardboard concrete-forming tube, 8" diameter, 1' length**
» **Power drill with ³⁄₁₆" and ¼" drill bits**
» **Phillips head screwdriver**
» **Fine tip permanent marker**
» **Clamps, 4" (2)**
» **Leather gloves and safety goggles**

Zakariya Al-Razi, the Secret of Secrets, and the
Invention of the Modern Still

Build your own time-tested rig for distilling essences or ethanol

Written by William Gurstelle • Illustrated by Peter Strain

Distillation is a purification technique that every chemist knows well. Since different chemical compounds boil at different temperatures, chemists can separate one desired substance from another through selective heating. We use distilling equipment to make everything from distilled water to ethanol to gasoline.

WILLIAM GURSTELLE's new book series *Remaking History*, based on this magazine column, is available in the Maker Shed, makershed.com.

Although distillation has been around since ancient times, the modern process traces its roots to medieval Islamic alchemists. Zakariya Al-Razi was foremost among them. A tenth-century Persian alchemist, physician, and philosopher, Razi is considered by many historians to have devised and put forward in writing the foundations of modern chemical distillation.

In his manuscript with the rather thrilling name *The Book of the Secret of Secrets*, Razi (also known as Rhazes or Rasis in the West) lays out the equipment, chemicals, and techniques involved in efficiently distilling a great variety of substances including kerosene, alcohol, and essential plant oils. While others wrote about distillation before him, it is Razi who first describes using the three main components of modern distilling equipment: the *qar'a* (boiler), the *anbiq* or alembic (distillation head) and the *qabila* (the liquid receiver.)

In this issue's Remaking History project, you'll follow in Razi's footsteps during the Islamic golden age and fabricate a working copper alembic pot still, capable of distilling many substances. With this equipment, you can make purified water, essential oils such as rose water and lavender oil, and even ethanol (alcohol) — which Razi also discovered!

1. CUT OUT THE PIECES

Use the ruler, metal scribe, and flat dividers to lay out cutting lines on the copper sheet for all the rectangular pieces, as shown in Figure Ⓐ, the cutting diagram. Don gloves and use the aviation snips to cut the pieces. Sand all the edges smooth.

CAUTION: Sheet metal work and soldering require goggles for eyes and gloves for hands. Burns and cuts may occur if proper safety measures are not followed.

Use the protractor, ruler, dividers, and scribe to lay out the boiler cape shown in the cutting diagram and Figure Ⓑ. Use the snips to cut the boiler cape. Sand the edges smooth.

In similar fashion, lay out the alembic cape as shown in the cutting diagram. Cut out and sand the edges.

2. MAKE THE BOILER

Using the cardboard tube as a form, bend the 8"×24" sheet into a cylinder. Remove the tube, overlap the ends by ½", and use the clamps to hold the copper sheet while you do the next two steps (Figure Ⓒ).

Starting ½" from the bottom, drill ³⁄₁₆" holes every 1" along the copper sheet, ³⁄₁₆" in from the edge.

Insert the #8 machine screws into the holes (Figure Ⓓ). Tighten the nuts on the inside, and then remove the holding clamps.

Flux the overlapping seam. Use the propane torch to heat the metal and when sufficiently hot, solder the joint. For a demonstration of how to solder a copper boiler, check out youtu.be/co5qlvGkQIs.

Center the copper cylinder on the 8" copper square. Trace the cylinder outline onto the square. Measure and mark another circle ⅛" inboard from the outline. Use the snips to cut out the circle of the outlined cylinder (cut on the outer line, not the inner line).

Starting from the scribed inner line, use the vise-grip pliers to bend a solder lip into the copper (Figure Ⓔ). It's best to do this in two passes so as not overstress the copper as you bend.

Insert the bottom into the cylinder. Use the pliers to adjust the solder lip on the bottom so it fits into the cylinder without gaps (Figure Ⓕ).

Flux the seam and solder, again using the pliers to close any gaps (Figure Ⓖ).

Carefully bend the boiler cape into the shape shown in Figure Ⓗ with the edges overlapping by ½". It may take some gentle prodding to make the cape completely round. When you're satisfied with the shape, use the vise-grips to hold the cape in place while you complete the next step.

Drill three ³⁄₁₆" holes in the boiler cape, ³⁄₁₆" in from the edge. Insert #8 machine screws and nuts. Flux the seam and solder.

Cutting Diagrams (not to scale)

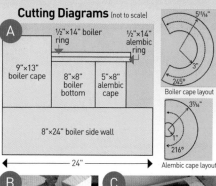

Ⓐ
½"×14" boiler ring
½"×14" alembic ring
9"×13" boiler cape
8"×8" boiler bottom
5"×8" alembic cape
8"×24" boiler side wall
24"

5¹¹⁄₁₆"
245°
3"
Boiler cape layout

3⁷⁄₁₆"
1"
216°
Alembic cape layout

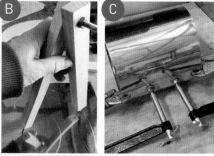

Ⓑ   Ⓒ

Ⓓ

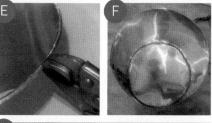

Ⓔ   Ⓕ

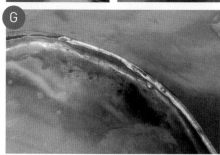

Ⓖ

Ⓗ

William Gurstelle

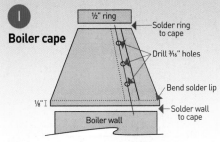

Boiler cape

½" ring
Solder ring to cape
Drill ³⁄₁₆" holes
Bend solder lip
⅛" I
Solder wall to cape
Boiler wall

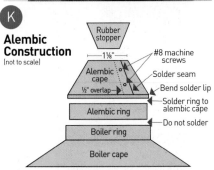

Alembic Construction
(not to scale)

Rubber stopper
1⅛"
#8 machine screws
Alembic cape
Solder seam
Bend solder lip
½" overlap
Solder ring to alembic cape
Alembic ring
Do not solder
Boiler ring
Boiler cape

William Gurstelle

Place the cape on the boiler cylinder wall. Mark a ⅛" overhang from the boiler wall on the cape and use the tin snips to cut and remove the excess. Use the vise-grips to bend a solder lip parallel to the boiler wall (Figure I).

Place the boiler cape over the cylinder wall. Use a hammer to gently tap the copper so there are no large gaps between the cape and the boiler wall. Flux and solder the wall and boiler cape together.

Carefully bend a ½"×14" inch strip into a ring that fits atop the upper opening of the boiler cape. Allow for a ½" overlap and cut off the remainder. Flux and solder the ring.

Place the ring atop the upper opening of the boiler cape. Flux and solder the ring to the cape as shown in Figure J.

3. MAKE THE ALEMBIC

The alembic detaches from the boiler to facilitate cleaning the inside of the still. The two pieces connect via two close-fitting rings, one inside the other. One ring you've already soldered to the boiler; the other you'll solder to the alembic. Then the rings simply slide over each other; this joint is not soldered (Figure K).

Bend the alembic cape into a truncated cone and overlap the edges by ½". The top opening should measure 1⅛". If not, readjust the cone. It may take some gentle prodding to make the cape round and symmetrical.

NOTE: If necessary, you can use the snips to cut the opening to 1⅛".

Use the vise-grips to hold the cape in place while you drill two ³⁄₁₆" holes, ³⁄₁₆" in from the edge and ½" from the top and bottom edges. Fasten with the remaining #8 machine screws and nuts. Solder the seam.

Carefully bend the other ½"×14" strip into a ring that fits snugly inside the ring on the boiler cape. Allow for a ½" overlap and cut off the remainder and solder the ring.

Place the alembic cape atop the ring and mark the circumference with a marker. Use the vise-grips to bend a solder lip; you may find that cutting a few slits at intervals makes it easier to bend (Figure L). Use the vise-grips to press out any large gaps, then solder the ring to the alembic cape. Figure M shows the finished alembic and boiler.

4. ADD A CONDENSER COIL

Drill a ¼" hole through a #7-sized silicone stopper. Insert one end of the ¼" soft copper tubing through the hole and bend the rest of the tubing into a gentle spiral.

Insert the stopper into the mouth of the alembic (Figure N). If necessary, use the vise-grips and a file to shape the alembic opening so the stopper fits snugly.

5. TEST FOR LEAKS

Fill the still with water and test for leaks. Fix any leaks by resoldering, or use high-temperature sealant.

Congratulations! Your still is complete.

USE IT

Your pot still has a capacity of a bit more than 1 gallon, which is perfect for many small distillation projects. It's suitable for making ultra-pure water, essential oils such as rose oil and lavender essence, and small batches of ethanol fuel. Depending on what you choose to distill, you may need to place the copper condensing coil inside a bucket of cold water in order to condense the distillate properly.

CAUTION: Plumbing solder melts at about 400°F. Use an electric hotplate; not open flames. Never heat the still without liquid inside, as the solder joints may melt.

There are many excellent books that provide instructions on how to use a still like this to produce essential oils, purified water, or alcohol. Read and follow those instructions, mind your local laws, and enjoy using the stuff you make. ●

GIVE A GIFT.
ONE YEAR ONLY $39.99.
Make:

GIFT FROM

NAME _____ (PLEASE PRINT)

ADDRESS/APT. _____

CITY/STATE/ ZIP _____

COUNTRY _____

EMAIL ADDRESS (required for order confirmation)

GIFT TO

NAME _____ (PLEASE PRINT)

ADDRESS/APT. _____

CITY/STATE/ ZIP _____

COUNTRY _____

EMAIL ADDRESS (required for access to digital edition)

☐ Please send me my own subscription of Make: 1 year for $39.99.

We'll send a card announcing your gift. Make: currently publishes 6 issues annually. Occasional double issues may count as 2 of the annual 6 issues. Allow 4-6 weeks for delivery of your first issue. For Canada, add $9 US funds only. For orders outside the US and Canada, add $15 US funds only. Access your digital edition after receipt of payment at make-digital.com.

484GS1

BUSINESS REPLY MAIL

FIRST-CLASS MAIL   PERMIT NO. 865   NORTH HOLLYWOOD, CA

POSTAGE WILL BE PAID BY ADDRESSEE

Make:

PO BOX 17046
NORTH HOLLYWOOD CA 91615-9186

Hands-Free Photography

Rig a DIY foot-switch shutter remote and give yourself a hand — or two — in documenting your project steps **Written by Becky Stern**

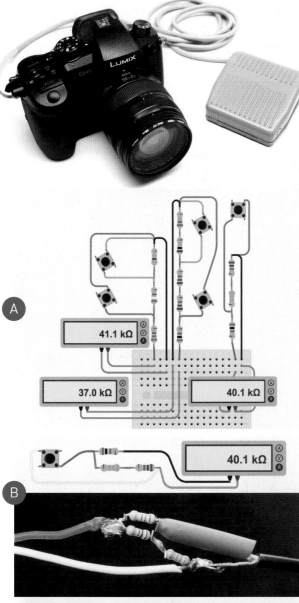

I do a lot of tabletop overhead photography featuring both of my hands, so a foot pedal shutter remote is an absolute must-have! Canon remotes are straightforward to DIY, but the remote for the Panasonic/Lumix GH5 (my camera) has a few resistors in it, which makes it a bit more involved.

I looked it up at doc-diy.net/photo/remote_pinout, and sure enough, the switch contact is held high at about 41.1 kilohms (41.1K), and the shutter triggers when the switch brings it down to about 2.2K. This isn't hard to re-create: Resistor values add up when put in series, and my experiments show you can successfully deviate a bit on the resistor values (try what you have that's close).

The circuit on the left in Figure Ⓐ is the ideal, 41.1K configuration (with the addition of a "half press" focus button which I did not include in my build), but I didn't have those exact resistors (2.2K, 2.9K, and 36K) in my collection. The center and rightmost circuits are my successful breadboard experiments. I chose to build the one on the right: 2K + 33K + 5.1K = 40.1K.

Strip 2" of insulation off the foot switch cable and the male 2.5mm TRRS cable, and check continuity with your multimeter to identify all the wires. Trim off unneeded wires — the GH5 remote uses only the sleeve and ring 2, not ring 1 or the tip.

Then solder the circuit together: the 2K resistor to the remote's positive wire (black); the foot switch wires to the 2K and the remote's negative wire (copper); and finally the 33K and 5.1K resistors in series across the gap as shown in Figure Ⓑ. Don't forget to add a large piece of heat-shrink tubing to your switch cable before you begin soldering!

Test the circuit to make sure the switch triggers your camera's shutter. The first time I built this, I mixed up the switch wires and had to make a fix. If the shutter remote is working, shrink the tubing around your circuit to seal it up (Figure Ⓒ).

What hands-free shots will you take with your own foot pedal shutter remote? ⊘

TIME REQUIRED:
1–2 Hours

DIFFICULTY:
Easy

COST:
$10–$20

MATERIALS
» **Foot pedal switch** Adafruit #423, Amazon #B01D8EHD1S, or Aliexpress #MKYDT1-201
» **Four-conductor (TRRS) 2.5mm male connector** I cut mine from a micro audio adapter cable, Amazon #B00FHBXL94. *TRRS* describes the four contacts (tip, ring 1, ring 2, sleeve).
» **Resistors: 2.2kΩ or 2kΩ (1), plus others to total ~38.9kΩ** I used 2kΩ, 33kΩ, and 5.1kΩ, but other combinations will work.
» **Heat-shrink tubing**

TOOLS
» **Multimeter**
» **Heat gun or lighter**
» **Eye protection**
» **Soldering iron and solder**
» **Wire strippers**
» **Flush cutters**
» **Helping third hand tool**

Becky Stern, Tinkercad Circuits

BECKY STERN
is a content creator at Instructables and the author of hundreds of tutorials, from micro-controllers to knitting. Previously, she was a video producer for *Make:* and director of wearable electronics at Adafruit. She also wrote "If This Then That" in this issue (page 28).

[+] Learn more at instructables.com/id/GH5-Foot-Pedal-Shutter-Remote.

Altered Apparel

Written and photographed by Lisa Mecham

Add sass to your simple sweater with ribbon and grommets

TIME REQUIRED:
1 Hour

DIFFICULTY:
Easy/Intermediate

COST:
$5–$7

MATERIALS
» **Sweater**
» **Polyester fabric** for lining
» **Grommet set,** ½" such as Dritz #44389
» **Satin ribbon spool,** ½"

TOOLS
» **Fabric scissors**
» **Tailor's chalk and ruler**
» **Basic sewing machine**
» **Hammer**
» **Dressmaker pins**

LISA MECHAM spends her days running a DIY fashion blog and her nights completing messy projects with her four kids. You can find more of her refashion and other DIY projects on her website, CreativeFashionBlog.com

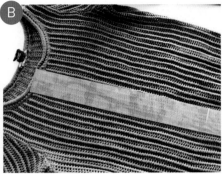

Ribbon really dresses up a basic knit sweater, and, unlike the bare-shoulder versions, a lace-up sweater will still help you stay warm. Here's how to make your own by repurposing one of your pullovers.

1. CREATE A LINER

Start out by cutting a small rectangle of fabric large enough to cover the neckline of your sweater. Lay the rectangular fabric piece over your sweater and use scissors to cut along the neckline so they match (Figure A). Remove the lining and set aside.

> **TIP:** Be sure your lining piece of fabric is 2" longer than your lace-up section of the sweater. This will help it all lay flat once you are done.

2. MEASURE THE LACE-UP SECTION

Decide how deep you want the lace-up part of your sweater to go. Use tailor's chalk to mark the line down the front of your sweater (Figure B). This will be used as a guide to be sure the fabric doesn't fray while you're working with it.

3. SEW AND CUT SWEATER

Use your sewing machine to stitch a line starting from the center of your neckline (Figure C), down the right side of the line, across the end of your tailors chalk line, and back up the left side of the line to create a long, skinny rectangle. This makes it easy to cut open a chunky knit sweater without the yarns and threads coming undone.

Now cut along the line you have drawn to open your sweater (Figure D).

4. ATTACH LINER

Lay your fabric lining over the front of your sweater again (right sides facing together) and pin into place along the neckline. Cut a

line through the rectangular fabric as well to match the cut on your sweater. Again, make sure your lining drops 2" below the cut on the front of your sweater.

Head back to your sewing machine and sew the lining to your sweater (Figure E). Start at a corner of the neckline and continue all the way down the slit, up the other side, and finish on the other side of the neckline. Flip the lining inside your sweater.

5. PLAY WITH GROMMETS!

Starting from the top of the neckline, I measured 1" down and placed my first grommet (Figure F). From there I continued punching grommets every 2" and setting them with a hammer (Figure G). You can fold your sweater in half and use dressmaker pins to mark where to place matching grommets on the other side.

> **TIP:** A grommet kit is the easiest way to get all the tools you'll need for this project without wasting your money on things you don't. The set I used is less than $9 and includes the right size grommet setter and enough grommets to complete your project.

Now lace it up with some pretty ribbon (Figure H) and you are all set! ✪

FABRIC FRIENDLY

While there are lots of Pinterest tutorials on how to refashion a basic cotton sweater, I wanted to create one that walked you through some simple hacks that make it possible to work with more complicated fabrics like chunky knits or loose weave. This way, no matter what textile you're dealing with, you can always create the finished look you want.

Ballpoint Penball!

Hack a click pen to make a mini pinball game

Written, photographed, and illustrated by Bob Knetzger

TIME REQUIRED:
A Weekend

DIFFICULTY:
Easy

COST:
$10–$15

YOU WILL NEED

- » A click pen (free!)
- » A marble
- » Lightweight chipboard or thin cardboard
- » Styrene sheet from a craft store
- » Pencil, hobby knife, and straightedge
- » Drill with ⅛" bit
- » Glue
- » Solvent bottle with MEK solvent

BOB KNETZGER is a designer/inventor/ musician whose award-winning toys have been featured on *The Tonight Show, Nightline,* and *Good Morning America.* He is the author of *Make: Fun!,* available at makershed.com and fine bookstores everywhere.

They're ubiquitous: those clicking ballpoint pens. Find them at the bank, the doctor's office, or almost any place where forms and signatures are needed. Go grab a freebie pen and make this fun bagatelle: the Ballpoint Penball game!

MAKE IT

For the ball launcher, use the ink cartridge as the plunger, the button as the tip, and the compression spring for the launching force. Reassemble the parts as shown, and discard the little cam and the barrel (Figures **A** and **B**). The cap holds things together, and you'll use the clip to help mount it to the game.

Make the game parts from styrene sheet, following the template at makezine.com/go/ballpoint-penball. To cut the pieces, score the plastic along a straightedge with the knife (Figure **C**), then bend it backward. It will snap cleanly along the scored line (Figure **D**). Cut out the playfield and plenty of rail strips.

Lay out the rail pattern onto the playfield

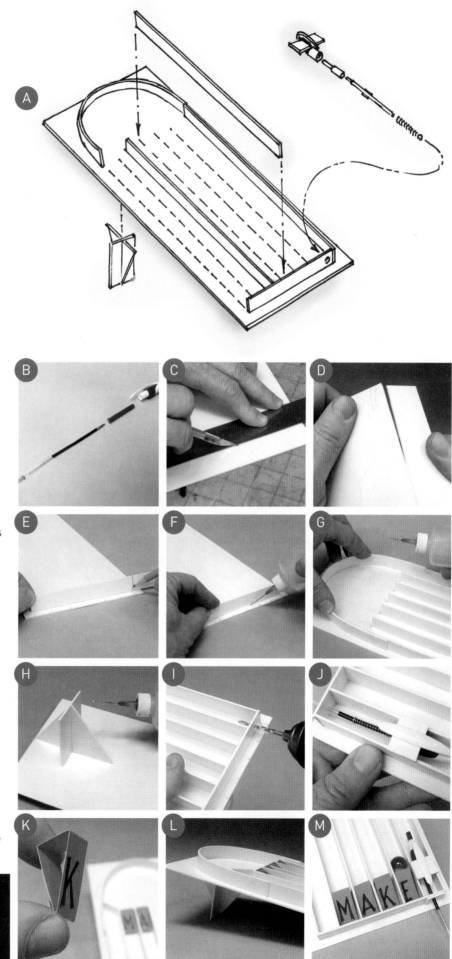

with light pencil lines. Hold a rail piece in position, mark the length needed (Figure E), then score and snap. Hold the trimmed rail piece back in place and, using the solvent bottle, apply a drop of MEK directly to the joint (Figure F). The MEK will flow along the seam. Carefully continue to hold the rail in position while the solvent bond hardens in just a few seconds.

Keep adding rails as you go (Figure G). The corner joints will provide strength. Check the spacing between the rails to ensure your marble rolls through freely. Last, add the stand on the back to give the playfield an angle (Figure H).

To mount the launcher, glue the pen clip to a piece of styrene strip. Drill a small ⅛" hole in the bottom rail (Figure I) so the ink cartridge can pass through — but not the spring. Slide the launcher assembly into position (Figure J), then move it down just enough so that the spring is slightly preloaded and still leaving plenty of travel for launching. When you've found the best position, solvent bond the cross piece to the rails. When the bond has fully set, test-fire the launcher. If needed, break the launcher free, reposition, and re-glue. By carefully pulling down on the ink cart "plunger" you should be able to regulate the shooting force and aim the marble into various tracks at the top, just like a real pinball game.

Make the four MAKE target sleds from lightweight chipboard (Figure K). Trim and check: they should slide easily down the playfield tracks when pushed downhill by the rolling marble (Figures L and M).

LET'S PLAY
Set all the targets at the top of their tracks and load the marble. For a simple one-player game try to score all four targets with the fewest number of shots — a perfect game is just four balls. Or, up the tension and play against time: How quickly can you score them all? For a two-player game, take turns shooting, but aim only for your targets (one player is "vowels," the other is "consonants," kind of like stripes and solids in pool). Make up your own game variations, too! ◗

[+] Find the template, see video, and get more details online at makezine.com/go/ballpoint-penball.

Corinne Elicone Insta: @Soundaffects

KAT CHAPMAN is a master gardener with University of California Cooperative Extension, San Diego County. She has volunteered with community and school gardens since 2000. mastergardenersd.org

Squarely Rooted

Use a grid to get big gardening results in small spaces Written by Kat Chapman

Home vegetable gardens are definitely back in vogue, and square foot gardening (SFG) is an interesting approach that lets people maximize their veggies regardless of the garden space they have. With the SFG method, you partition a raised bed into 12" squares (hence the name) with wood or string overlays, then grow a specific amount of one type of vegetable in each section, based upon each veggie's size and spacing needs — 16 parsnips, 9 green onions, 4 chives, and so on. It's a simple concept that allows novice gardeners to get the most out of their space.

COMPLEMENTARY CROPS

The plants share the same soil, and can be optimized for compatibility following recommended combinations. (The same holds true for non-SFG raised gardens.)

Tomatoes, eggplant, kale, peppers, tomatillos, and potatoes (one each per square) will all complement each other and have the same water requirements.

BETTER BEDS

Using raised beds provides many advantages:
» They don't take up a lot of space
» You use your own nutrient-rich soil
» Because the garden is contained, water consumption is much less, and excess water easily drains away from the plants and prevents rotting
» Bugs, slugs, and critters will have a harder time munching on your veggies
» Weeds are less likely to get a hold and much easier to pull out

When you construct your raised bed, line the bottom of the bed with hardware cloth

(not chicken wire) to keep gophers out. A simple drip irrigation system will reduce the time spent watering, and wet the soil not the leaves.

GARDEN GUIDANCE

There are a wealth of apps and information available online to help you create your exact garden setup. Check out growveg. com/guides/categories/square-foot-gardening for guides on building planter boxes and lists of vegetable spacing to use; vegetablegardeningonline.com offers a drag-and-drop tool to help you design your layout. You can find a good list of compatible plants at growingplaces.org/wp-content/uploads/Raised-Bed-Garden-Methods.pdf. Good luck! ◗

1+2+3 Straw Oboe

Written by Matt Stultz • Photographed by Enrico Spadoni

Avoid the begging for screens and the whines of "I'm bored" with a bit of colorful creativity. Craft simple wind instruments from drinking straws and strike up a summertime band.

1. CREASE THE STRAW

Pinch the end of the straw tightly together, creasing the sides from the tip to about 1 inch down.

2. SHAPE THE END

Trim the tip of the straw on both sides along the length of the pinched area. You want a good point at the end of the straw.

3. CUT HOLES

About a third of the way down the straw, use your scissors, wire or other cutters to make two angled cuts forming a diamond-shaped hole in the straw. Add two more holes spaced evenly down the length of the straw.

PLAY IT

Place the pointed end of the straw in your mouth. Push down with your lips to help squeeze the straw and blow hard to make the pointy tips vibrate together — like the reed in a real oboe.

Cover and uncover the holes to play different notes.

GOING FURTHER

Try adding more holes to your straw. How about a series of straws, but instead of poking holes, you make them all different lengths? Try using a piano or guitar tuner on a smartphone to make a straw oboe that hits a note perfectly. ⊘

TIME REQUIRED:
2–5 Minutes

DIFFICULTY:
Ridiculously Easy

COST:
Pennies

MATERIALS
» Drinking straw

TOOLS
» Scissors with sharp tips, or wire cutters

MATT STULTZ
is *Make:*'s digital fabrication editor. He is the founder of 3DPPVD, Ocean State Maker Mill, and HackPittsburgh.

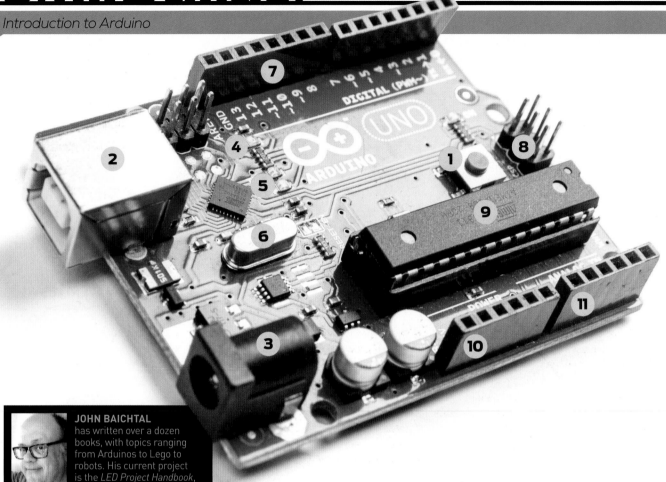

JOHN BAICHTAL
has written over a dozen books, with topics ranging from Arduinos to Lego to robots. His current project is the *LED Project Handbook*, forthcoming from No Starch.

ARDUINO TOUR

1. Reset button — This button restarts the Arduino sketch from the beginning.

2. USB-B socket — This socket accepts a USB-B cord.

3. Barrel plug — Plug a power cord with a 2.1mm barrel into this power socket.

4. Built-in LED — Want to test out the Arduino? The easiest way is to blink this LED, see "Program the Device."

5. TX/RX LEDs — These LEDs flash to show that data is flowing. You'll most often see this when uploading to the Arduino.

6. Timing crystal — This 16Hz crystal helps the Arduino keep track of time.

7. Digital pins — These pins control LEDs, servomotors, and other components.

8. ICSP header — This 6-pin header allows the Arduino's bootloader to be reflashed, or reprogrammed, without removing the board from a hypothetical circuit.

9. ATmega328P — The brains of the Arduino, the 328P is a microcontroller that controls the digital pins and reads data from the analog pins.

10. Power pins — These pins deliver 3.3V or 5V plus ground and a few other bells and whistles.

11. Analog In pins — The Analog In pins are typically used to take readings from sensors.

Get to Know
Arduino

Here's a crash course on the popular microcontroller Written by John Baichtal

IF YOU HAVEN'T JUMPED INTO THE ARDUINO WORLD YET, LET'S GET YOU UP TO SPEED. Arduinos are microcontrollers, credit card-sized computers that listen to sensors, power lights, and control motors, following a series of programmed instructions. They're easy to use and cost hardly anything by technology standards.

NUMERO UNO

The Uno isn't the fastest, the newest, or the most powerful Arduino around. However, it's fine for most projects, and there are tons of code examples and tutorials available for it, not to mention add-on sensors and components galore. For these reasons, the Uno remains popular, even among tinkerers who have moved on to bigger and better models.

PUT GPIOS TO WORK

Arduinos interact with the world with the help of their general-purpose in/out (GPIO) pins, analog and digital. Digital pins turn on and off to control 5-volt signals, allowing you to power LEDs, trigger motors, and so

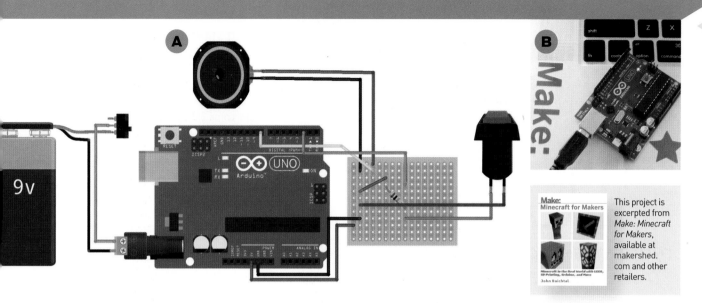

on. Digital pins can also create dimming effects by turning components on and off very rapidly.

By contrast, analog pins take a range of readings from sensors. For instance, a light sensor connected to an analog pin typically delivers a reading from 0 to 1027 depending on the brightness of the detected light.

Figure **A** is a rendering showing a simple Arduino project. It sounds a buzzer when the button is pressed. The yellow wire in the diagram connects the buzzer to Pin 8 and the orange wire connects the button to Pin 2. The red and black wires provide power and ground.

POWERING THE UNO

There are three ways to power the Arduino. The first and easiest consists of plugging the board into a computer via a USB cable. This not only powers the Arduino but also serves as a data connection so you can program it. Requiring a computer is one limitation of this method, though a wall-plug adapter or USB-equipped battery can work as well. Figure **B** shows the Arduino powered up via USB.

There is also a barrel-plug that accommodates battery packs and "wall wart"-style power supplies. A good example of such a power supply is the 9V wall adapter from SparkFun (part #298), though you can use a 9V battery with the right plug as well.

Finally, the pin marked Vin near Analog 0 can be used to power the board if you want a soldered connection instead of a plug.

DOWNLOAD SOFTWARE

To program the Arduino, you'll need to download the development environment, the Arduino IDE. You can find it at arduino.cc under Software. The environment works on Windows, Linux, and Macintosh computers but the directions differ for each platform, so I suggest you peruse their installation instructions.

PROGRAM THE DEVICE

Arduino programs are called sketches, but they work using the same universal programming techniques that other languages offer, making the process very simple and understandable if you've ever done any programming. Even if you haven't, the syntax is straightforward once you get used to it.

Let's examine Blink, one of the simplest sketches as well as the one newbies typically learn first. Blink does what you expect: it makes an LED blink. More specifically, it makes the built-in LED on the board blink, one second on and one second off.

Go into your Arduino software and select the Blink sketch using File → Examples → Basics. Open it up and take a look. The first part is **setup(),** which does just what you'd expect — it gets the program elements ready to go. **setup()** runs once when the Arduino is powered up; as long as the power isn't shut off and restored, **setup()** won't rerun.

Here is the setup for Blink:

```
void setup() {
pinMode(LED_BUILTIN, OUTPUT);
}
```

As you can see, **setup()** encompasses the area within the curly braces, which consists of one command: initializing the built-in LED's pin as an output pin.

UPLOAD SKETCHES

Once you have a sketch open on your screen, follow these steps:
1. Select Arduino/Genuino Uno from the Tools → Boards drop-down menu.
2. Select a serial port from Tools → Port. Choose any one that works.
3. Before you upload, you might want to check the code by choosing Sketch → Verify/Compile, which will sniff out any errors without trying to upload.
4. Double-check that the USB cable is connected.
5. Click Sketch → Upload to send the sketch to the Arduino.

NEXT STEPS

The best way to learn about Arduino is to explore the example sketches found with the Arduino software. Choosing File → Examples in the Arduino software takes you to a collection illustrating how to do pretty much anything with the boards. Additionally, the Arduino Playground (playground.ardino.cc) offers help and suggestions for new and experienced programmers. ⊘

Tech Tunes

Create music from floppy drive sounds with *a Java application and Arduino sketch*

Written and photographed by Sam Archer

SAM ARCHER Software engineer by day, gamer and tech hobbyist by night.

TIME REQUIRED:
1–2 Hours

DIFFICULTY:
Intermediate

COST:
$20–$50

MATERIALS

» **Floppy drive, 3.5"** at least 1 but as many as 8 or more
» **Arduino Uno microcontroller board** or equivalent
» **Power supply, AT or ATX** though any 5V 2A+ supply will do
» **Floppy drive ribbon cable, 34 pin IDC connector**
  —or—
  Jumper wires, male to female (3) and female to female (1)
» **Breadboard and some extra wire (optional)** but might make organization easier

TOOLS

» **Wire cutters**
» **Wire strippers**
» **Utility or X-Acto knife (optional)** makes separating ribbon cable wires easier
» **PC** capable of running the Arduino IDE (arduino.cc/en/Main/Software) and Java
» **Arduino IDE** with TimerOne library installed

IT STARTED AS A BET: THE "PHANTOM OF THE FLOPPERA" VIDEO WAS MAKING THE ROUNDS AT WORK and a co-worker was sure that the video was fake.

"They just dubbed audio over a video of some old computer," he said.

"Nonsense," I said. "It's not even that hard! I bet I could go home tonight and do it!"

After many hours and a lot of trial and error, my salvaged floppy drive eked out the first few bars of "Ode to Joy," only a little bit out of tune.

Floppy disks have several "tracks" of information on their circular media, and the drives that read them have a head that can step one track at a time across their surface. The position of the head is controlled by a stepper motor, and floppy drives have a convenient interface that moves the head one track at a time with a signal pulse to the right pin. Pulse this signal at a specific frequency and the drive's vibrations give off sound like a speaker driven by a square wave.

After getting a glimpse of the potential musical floppy drives had to offer, I dug out an Arduino, mooched a grocery bag full of floppy drives off a friend, and set to work making beautiful (or at least recognizable) music. The result? A Java application and Arduino sketch that, as this article will show you, can be used to make floppy drives sing.

Hardware Assembly

First you'll need your Arduino, floppy drive(s), power supply, and the floppy drive ribbon cable or jumper wires (Figure Ⓐ). If you're using a ribbon cable, cut off one end of the cable leaving enough cable to work with (some cables have a twist in them; avoid the twist side to make it easier to follow the wires).

The floppy drive pins are arranged in pairs: odd pins are grounded and even are data. Signals are sent to the data pins by grounding them. There are three data pins that we're interested in:
» **12 (or 14) — The "Drive Select B (or A)" pin**
» **18 — The "Direction" pin**
» **20 — The "Step" pin**

The Drive Select pin is used to activate the drive so it will respond to signals. When the drive is powered and activated, its LED indicator will light up. The Direction pin determines if the drive's head moves forward or backward. And each time the Step pin transitions from ungrounded (floating or at 5V) to grounded, the drive will move the head one step in the direction of travel.

If you're using a ribbon cable, separate the wires as shown in Figure B; if you're using jumper wires, connect them to the pins on the floppy drive directly. The Drive Select wire should be connected to its matching ground (either 12–11 or 14–13). Some drives will work with either pair connected, but some will only work with one or the other, so you may need to experiment with your drive to find out which pair of connected pins causes the LED on the drive to turn on (in these photos it's 12–11).

The Step wire should be connected to pin 2 on the Arduino, the Direction wire to pin 3, and it's very important that one of the grounded wires is connected to a GND pin (Figure C) on the Arduino (without this, the Arduino and drive may disagree on "grounded" and signaling may not work). If you have more than one floppy drive, the additional drives' Step wires should be connected to the even-numbered Arduino pins (up through A2 for eight drives), and the Direction wires should be connected to odd pins (up through A3).

Once the drive is wired to your Arduino, it's time to attach the power supply. If you're using an ATX power supply, use the Molex-type floppy drive connector; if you're using a 5V power supply, the rightmost pin on the floppy drive's power connector should be +5V and the second-rightmost should be ground (the red wire on the standard power connector is +5V, the black is ground). If you're using an ATX power supply, you'll also need to short the green and black wires on the large connector to signal the power supply to turn on (Figure D).

Once powered, the drive's LED should light (Figure E). If it doesn't, verify your wiring and try switching between the A and B Drive Select pins.

Arduino Sketch

Download and install the Arduino IDE. Open it up and go to Sketch → Include Library → Manage Libraries, and search for and install the *TimerOne* library. Close the Arduino IDE and plug your Arduino Uno into your computer's USB port. It may take a moment for the drivers to initialize, but installing the Arduino IDE should have taken care of driver installation as well.

Download the *MoppyArduino* zip file from github.com/SammyIAm/Moppy2/releases, unzip it somewhere convenient, and open the *Moppy.ino* sketch file in the Arduino IDE. The IDE should have automatically detected that you have your Arduino plugged in, but if not you may need to select the appropriate board type and port in the Tools menu.

Click the Upload button at the top of the IDE. The IDE should successfully compile and upload the sketch to your Arduino. Pay close attention to the black console at the bottom of the window to make sure that there are no errors in red text.

If the upload was successful and your power supply is on, the first floppy drive you connected should play a little four-note startup sound anytime the Arduino is reset.

Control Software

Now download the *MoppyControlGUI* zip file from the same GitHub page, unzip it somewhere convenient, and run the appropriate *MoppyControlGUI* executable from the *bin* folder (e.g. **.bat* if you're on Windows). Select the checkbox under Network Bridges that corresponds to the serial port your Arduino is associated with; the Arduino should reset and will display as a device on the right side of the window.

Click "Load File" to load one of the included MIDIs from the *samplesongs* folder, then press the play button!

GOING FURTHER

Now that you've got at least one musical floppy drive, consider adding more to your orchestra! Try mounting your drives on something aesthetically or acoustically pleasing (maybe with some RGB LEDs). Play around a bit with different mapping settings in the lower section of the *MoppyControlGUI* application, and try looking for some additional MIDI files on the internet to play.

If you're feeling especially crafty, this setup is adaptable for playing other stepper motor devices, or playing relays, for a computer-hardware percussion section. The software is all open source too, so any customizations are only a fork away. ⊘

LED is lit

MAYKU FORMBOX

$700 mayku.me

Makers have many tool options when it comes to creating a project but few provide such instant gratification as a vacuum former. The Mayku FormBox takes all the guesswork out of vacuum forming. Simply place one of their "Form" or "Cast" sheets in the machine, change the settings to those specified for your material, and wait for the timer to let you know it's ready to form.

Inside the box you get the FormBox itself, a selection of materials, a few objects that help teach you the basics of vacuum forming, and an adapter hose to hook up to your own vacuum. For the price I would love to see a built-in vacuum, but the system worked great with my home vacuum cleaner. My only real detraction of the FormBox is that it could use a few more inches in material size. I've built a couple vacuum formers in the past, but the Mayku FormBox showed me everything I did wrong with how right it is. *–Matt Stultz*

HOMERIGHT SPRAY SHELTER

$40 homeright.com/products/small-spray-shelter

Spray paints and finishes are easy ways to get color onto your workpieces, but if you work in a small shop, you'll no doubt find yourself struggling to control overspray to avoid making a mess.

The HomeRight Spray Shelter might be the solution you need. Available in two sizes, I've been using the smaller setup for a few months now, and it's made it a lot easier to get better spray finishes on my projects. From the small 12" diameter pouch, the Spray Shelter quickly and easily unfolds into a 35"×30"×39" spray booth. There's a large opening to easily get at all sides of your project, and the nylon material prevents paint or finishes from going where they shouldn't. It's a little tricky to fold back down when you're done, and I would have liked some attachment points to hang parts inside the booth, but overall, it's a fantastic addition to my workshop. *–Tyler Winegarner*

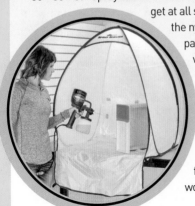

POKIT METER

$70 (Pre-Order Price) pokitmeter.com

A multimeter is vital for troubleshooting anything electrical, but most are too bulky to carry with you. Not the Pokit Meter. At just under 2" in diameter, it fits comfortably in a pants pocket or can be attached to a key chain or lanyard. The two test probes extend around 9", and a button quickly retracts them.

Connect the Pokit Meter to your smartphone via Bluetooth and things start getting really good. For metering, Pokit offers a voltmeter, ohmmeter, ammeter, and a thermometer. Pokit also features an oscilloscope and a datalogger, which allows you to track data over a period of up to 6 months, using sampling intervals from 1 second to a little over 2 hours.

Unfortunately, the curved probes are a little tricky to work with, and the Bluetooth connection means there's a bit of latency in your metering. Still, you're not likely to find a broader set of tools in a slicker package anywhere else. *–Tyler Winegarner*

SKILSAW SPT77WML-01 WORM DRIVE SAW

$200 skilsaw.com

At a crossroads after the batteries for my aging cordless tools crapped out, I'm now slowly replacing them all with corded versions that offer features, power, and longevity well beyond what the portable options provide. The first I've moved to is the classic worm-drive circular saw by Skilsaw. Its 15-amp motor seems to instantly hit its top speed with the click of the trigger, and unlike my old battery powered tools, it doesn't slow down when the blade starts passing through material. Geared for plywood and 2×4 lumber, I've sliced effortlessly through everything I've put in front of it.

The platform is thick and sturdy. Adjusting the blade height and angle requires little effort but holds solid when locked in place. And the motor's placement gives the tool an overall balance that I haven't felt with normal sidewinder-style circular saws. The cord hasn't been an issue either; my projects all happen in my garage or yard, close enough to an outlet for the cord (or a heavy-gauge extension) to easily reach. I look forward to decades of cutting with this now in my arsenal. –*Mike Senese*

MINTYPI KIT

$160 mintykit.com

As makers, especially those with any kind of love for retro gaming, we've seen a number of DIY Raspberry Pi-based handheld gaming consoles.

The MintyPi is a Raspberry Pi Zero-based portable emulator that runs RetroPie. It has a great little PCB that manages the connections to the buttons, display, speaker, and battery hardware. The project is a bit of a challenge (took me about 8 hours total) but very well documented and I had an absolute blast making it.

Admittedly my favorite part is using the finished product in public. Something about it being built into an Altoids tin makes me feel like the nerdiest version of James Bond imaginable. That ... or some guy who is really protective of his mints.
–*Andrew Stott*

KLEIN TOOLS TRADESMAN PRO ORGANIZER (BACKPACK)

$90 kleintools.com

Toolboxes are great but on the inside they often become disorganized piles of tools that you have to dig through to find the perfect screwdriver. The Klein Tools Tradesman Pro Organizer has 35 pockets, giving it a spot for every tool you need to carry. The rugged rubber bottom means you will never need to worry about where you set the bag down.

The pockets inside are set up for hand tools like screwdrivers, pliers, and wrenches, though I would love to have a bigger pocket or two inside for holding drills or other power tools. While not everyone needs a backpack full of tools, this is a great solution for getting your own implements back and forth to your makerspace or for anyone attending a trade school.
–*Matt Stultz*

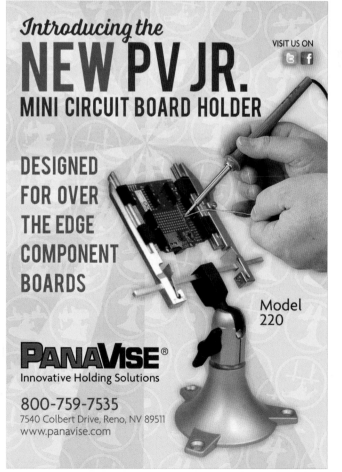

SHOW & TELL

Get inspired by some of our favorite entries to the most recent Maker Share contest

If you'd like to see your project in a future issue of *Make:* magazine submit your work to makershare.com/missions/mission-make!

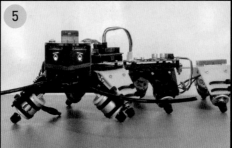

1 Inspired by Folker Stange's Thingiverse design, **Beth Sallay** built programmable Glockenspiels from repurposed thrift store xylophones. "I played around with the design in Adobe Illustrator. Going from ... clear acrylic to wood was interesting and took about two months of trial and error. Lots of testings and swearing," says Sallay. Her Glockenspiels were positively received by children and adults alike at Sallay's local Mini Maker Faire. makershare.com/projects/programmable-glockenspiel

2 With his Raspberry Pi Radio, **Kurt Lewchuk** converted a vintage radio to digitally behave as it did during the golden age while bypassing the original analog electronics. It's programmed to mimic the antiquated behavior as well as the vintage programming — adjusting the tuning knob lets him tune in various stations with AM noise (provided by WAV files) in between. "This is a conversation piece in my home," he says. makershare.com/projects/raspberry-pi-vintage-radio

3 After falling in love with the designs of 3D bathymetric charts, **Mike Niedbala** decided he wanted to create his own topographical map. "But I also wanted the map to mean something to me," Niedbala says. Settling on West Point, N.Y. (where he attended undergrad), Niedbala utilized the laser cutter in the makerspace at his grad school to make his dream into a reality. makershare.com/projects/wood-topographic-map

4 **Andre Ferreira** enjoys playing paragliding games, but struggled with finding a controller to get the most out of the experience. So he decided to make his own. "I had various concepts but ended up going for sliding potentiometer because they are cheap, compact, and readily available," Ferreira says. We love the overall design of the chassis, and the Arduino hidden within lets players reprogram the controller for plenty of other types of games. makershare.com/projects/diy-game-controller

5 **Cole Brauer**'s open source, Arduino-based, robotic centipede is perfect for teaching students the fundamentals of programming and electronics. "What sets Make-A-Pede apart from other DIY robots is that it is based on a segmented chassis design," Brauer says, "giving the robot a centipede-like motion and allowing the robot to be easily expanded. Segments are driven using pairs of motorized 'legs', and can carry a variety of different modules." makershare.com/projects/make-pede-0